U.S. Military Advisors in Vietnam

Uniforms and Equipment of

U.S. Military Advisors in Vietnam

1957-1972

Paul W. Miraldi

Schiffer Military History
Atglen, PA

ACKNOWLEDGMENTS

This book is the result of many individuals who gave their time and help willingly and freely, and gave me guidance in life. Without their assistance this book could not have been completed!

The following are those who deserve a special thanks and who made the effort possible: Kevin Lyles, Eric Halsey, Main Photographer, Author and Mentor, CSM Mike Martin, Mike Tucker, Steve Nelson, Tim Newark; Publisher "Military Illustrate," Reynold Cindrich, Dave George, and Kaiser Bills, Nick Friedrich, Kurt Wheeler, Jeff Warner, Arik Greenberg, Mr. and Mrs. Greenberg, Ben Singer, Justin Atwood, A.J. Allen, Jon Jones, Frank G. Benns; Ranger Co. C, Owen Thorton "the Mic," John O'Conner, John Conners, Glen Craigie, Gene Collins, Ed Condby, Sharon from Paul Gaudett Books, Alex Allan, Harlan Glenn, Jack McFerson, Jack Kelley, Larry Martin, Mike McDaniel, Bruce McIver, Robert Riley, Cyril Write, J.C. Fischer, Steve Watson, Sam Lee Pfister, Don Driftmire, Don Slinkard, Dave Williams, Nat Ward, Jack Frost, Carl Farren, Robert Parrish, Bob Chatt, Jay Camp, Dick Samms, the wonderful professionals at Target Chino film developing center, Dr. Sayles, and Dr. LeSueur at the University of La Verne, Mr. Clay Hess, Mr. Teft, the late Mr. Scherff, who inspired my love for History at Glendora High School, Ray and Ron Torres, Ed Davies, Gary McDonnald, Sgt. Donald Glover; 25th Infantry, First Sgt. Minor Swackhammer, and Joe, Amy, and Dale Lientz. Also my wife, Amber; Kirk and Barbara Skaggs for proofreading; Mr. and Mrs. Greenberg for proofreading; and my mother, Pat Pahel. Also a big thank you to Ian Robertson, Bob Biondi, and Peter Schiffer!

DEDICATION

This Book is dedicated to everyone who served as an advisor to the South Vietnamese Armed Forces, along with the brave South Vietnamese Forces.

Book Design by Ian Robertson.

Library of Congress Catalog Number: 00-107890

Printed in China.
ISBN: 0-7643-1183-2

We are interested in hearing from authors with book ideas on related topics.

Published by Schiffer Publishing Ltd.
4880 Lower Valley Road
Atglen, PA 19310
Phone: (610) 593-1777
FAX: (610) 593-2002
E-mail: Schifferbk@aol.com.
Visit our web site at: www.schifferbooks.com
Please write for a free catalog.
This book may be purchased from the publisher.
Please include $3.95 postage.
Try your bookstore first.

In Europe, Schiffer books are distributed by:
Bushwood Books
6 Marksbury Avenue
Kew Gardens
Surrey TW9 4JF
England
Phone: 44 (0) 20 8392-8585
FAX: 44 (0) 20 8392-9876
E-mail: Bushwd@aol.com.
Free postage in the UK. Europe: air mail at cost.
Try your bookstore first.

Contents

Foreword

In this second major title, Paul Miraldi has turned the focus of his attention to the advisory effort in Vietnam. While of a more complex nature than his previous subject, he has tackled it with his usual vigor and thoroughness. In this volume there are countless hitherto unpublished photographs from the collections of the veterans themselves. It is no small credit to him that Paul has gained the trust of this select group to the point where they are able to entrust their precious slides and photos to his care, not to mention the mercies of the U.S. Postal Service. We are thus presented with scores of previously unseen images that record both idiosyncrasies of uniforms peculiar to the advisors, as well as an insight into his life among the Vietnamese. It should be remembered that in many cases the informative content of the photograph is of greater importance than its quality; these are, after all, snapshots of otherwise busy men. In addition, Paul has drawn together another impressive collection of original uniforms and insignia. Much of this is of the utmost rarity, with a corresponding price tag on the militia market. Reproductions and fakes abound, so it is of great value to the new collector to have detailed reference of the real thing at his fingertips, which alone would be worth the book's price.

Kevin Lyles
Nov. 22, 1999

Introduction

The primary purpose of this book is to accurately show the appearance of U.S. Advisors, along with their units' Histories and Advisor duties. Also, to accurately show the uniforms, equipment, and personal items carried by the Co Van My (American Advisors) during the Vietnam War. This book will also show that you can "never say never, and never say always," that is, every advisor was different. I have found out over the years that anytime you start saying "All Advisors," or "No Advisor" wore this, you will be wrong every time. These re-creations are the exception, and are not always the rule. In other words, not everyone who served as an advisor looked like these re-creations. All photo recreations are the product of photos, or from 1st person oral explanation from the vets who wore them. Strict uniform policies in I Corps may not carry over to IV Corps. I have tried to show interesting uniforms and not recreate photos that we in the collecting field have seen a hundred times, over and over again. I have also corresponded with veterans to obtain new and exciting photos that no one has seen before. For example, the photos in Chapter 2 of the MAT team advisor. Mr. Don Driftmire was kind enough to come to my home with over 600 slides and a slide projector and allow me to make prints from his personal collection. I chose the 30 best that I thought I would want to see had I bought this book. And, finally, I put in information that I thought was interesting. This book is the product of years of exhausting research, trials, and errors, along with extensive buying of uniforms and gear, many hours on the phone to Kevin Lyles, Author, in England, correspondence by mail, looking at thousands of photos and books, talking and meeting with Vets, as well as comparing notes with other collectors throughout the country. Another name that needs to be mentioned is by all means the Guru of Vietnam Uniforms and Equipment, Mr. Shelby Stanton, literally the Godfather of Uniform Histories! Without Mr. Stanton's outstanding research and his books, US Army Uniforms of the Vietnam War, and US Army Uniforms of the Cold War, 1948-1973, this book would not have been possible, and many of us would still be in the dark about uniforms and equipment. Subsequently, Kevin Lyles' Vietnam: US Uniforms in Color Photographs, and his book on Infantry, withPhundreds of original unpublished photos is a must, and without it and Kevin's help this book could not have been made possible. The chapters are divided into Army, Navy, and Marine Corps, with the Army being subdivided into early MAAG Advisor, MACV Infantry and Armor advisor, Airborne advisor, and finally, ARVN Ranger advisor. This makes it easier for the collector to open the book to a specific chapter to find the information he wants. Each of the 6 chapters are shown in chronological order, and the figures are presented in chronological sequence so that the gradual development of uniforms, equipment, and weapons can be traced from year to year. In this way, it is possible to note at a glance when specific items were introduced into "service."[1] Each chapter will show the development of the items that were available, with close-ups to further help the Historian, collector, museum staff, etc., identify future items, or items they may already have. Chapter 7 is the shortest, having several photos of "personal items" carried by the advisors, as well as items he would have left back at his bunk. You may also note that some chapters are longer than others, and some Advisors may be represented in greater numbers than others, too. Please do not think this is due to a bias on the part of the author. Certain Advisors' uniforms were harder to obtain over the years, and some uniforms did not make it into my collection as well as unit histories, but rest assure that in the next book I will cover more of the units not represented in this volume. Due to time restraints and lack of items, many Advisors were not covered fully.

The "scope" of the book has been limited to U.S. Army, U.S. Navy, and U.S.M.C. Advisors, only. Unfortunately, I was unable to obtain enough information on size and location of any of the Air Force advisory elements, but not from lack of trying. In the second part of the advisor book I will definitely show as many of the Air Force's contributions to this period of history, along with the CAP Marines, other Navy advisors and artillery, MP advisors, and so on. If you were an Air Force advisor or any other advisor during the Vietnam War, please do not hesitate to contact me if you feel your specialty was not represented in this volume. I would be happy to speak with you and to make sure your unit is properly represented in the next volume of Advisors. Thank you for your patience and understanding. All letters can be addressed to me C/O Schiffer Publishing, Ltd., 4880 Lower Valley Road, Atglen, PA 19310. Paul W. Miraldi. This book is the second in a series I hope to publish in the coming years dedicated to the appearance of the American Fighting Man during the Vietnam War.

1

MAAG - Military Assistance Advisory Group 1955-1960

"The formal involvement of U.S. military personnel in Vietnam had begun in August 1950, when the first contingent of the U.S. Military Assistance Advisory Group (MAAG) arrived in Saigon to set up shop. The MAAG consisted of a joint team of Army, Air Force, and Navy personnel; its purpose was to provide the military supplies necessary to prevent expansion of communism in Indochina. Because the supplies were turned over directly to the French Expedition Corps, the MAAG members functioned in the early days primarily as a logistical accounting group with no real advisory role. The French adamantly refused to allow the Americans in MAAG to provide any training assistance to either themselves or their protégé, the Vietnamese National Army. The Vietnamese Navy, such as it was, was under the direct command of a French Officer, and by 1952 reports from American Naval officers in the MAAG were taking a pessimistic view of French claims to have established a Vietnamese Navy that would someday be able to function independently. This Navy consisted of the approximately 400 sailors serving under French officers and petty officers in the river forces, and a functioning training apparatus was nearly nonexistent."[2]

"American impotence in the advisory role continued throughout the next three and a half years, while the French and their Vietnamese allies fought against the determined forces of Ho Chi Mihn and his Communist Viet Mihn troops."[3]

"By May 1955, the U.S. contingent had reached 155 advisors. During this time, tensions developed between the Americans and the French serving in Vietnam. Part of the problem lay in disagreements over how the advisory mission was to be accomplished, but these disagreements were exacerbated by President Diem's preference for the Americans. The French were humiliated, for example, when the South Vietnamese Army abruptly converted its uniforms from French-style to a more American type and began saluting as the Americans did, rather than with the French open palm method. Too, American advisors apparently made no real effort to conceal their eagerness to take over from the French."[4]

"In the years immediately following the French departure, the American advisory commitment in Vietnam remained small. Despite the U.S.' refusal to sign the Geneva Accords in 1954, official government policy restricted the number of advisors to the 342-man limit imposed by the accords. When the French were in the final stages of withdrawal in 1956, the South Vietnamese government requested increased U.S. support, arguing that replacing the departing French with American advisors would be in keeping with the spirit of Geneva. Still concerned about the image of adhering to the limit, however, the U.S. administration sent a 350 man increase to Vietnam, not as an addition to the MAAG, but under the new auspices of a Temporary Equipment Recovery Mission (T.E.R.M.)"[5]

"In an attempt to counter the poor performance of the South Vietnamese troops, Lieutenant General Samuel T. Williams, then chief of the MAAG, had sent a letter to Admiral Harry D. Felt, Commander in Chief, Pacific (CINPAC), requesting a revocation on the long-standing restriction against allowing U.S. advisors to accompany their Vietnamese counterparts on combat missions.

U.S. Army Col. Nat Ward, III. Chief of Staff, MAAG, Saigon 1958-1960. Note that Col. Ward wears the Army Green Wool uniform. (Nat Ward collection)

Senior advisors' Conference Participants, Saigon, January 1960. (Nat Ward collection)

COAT, MAN'S, WOOL SERGE.
ARMY GREEN 44
Size 3 9 L
MERIT CLOTHING COMPANY, INC.
CONT. NO. DA-36-030-QM-8615-01-766-C-57
DATED 23 AUGUST 1956
Spec.Mil-C-1-13990(QM)-8615-01-766-C-57
STOCK NO. 8405-286-5158
PHILA Q.M. DEPOT – 100% WOOL

Tag from Coat, Man's, Wool Serge Army Green Shade 44. Note the date, 1956. This is the first year of the Class A Green Uniform. (Nat Ward Collection)

General Williams believed that the South Vietnamese failures had resulted from a lack of proper planning and aggressiveness, and he expected the presence of U.S. advisors to at least partially correct those problems. On 25 May 1959 Admiral Felt approved the request: American advisors were permitted to go along on operational missions, provided that they themselves not actively participate in actual combat. This caveat of nonparticipation often proved unrealistic when advisors found themselves under fire, and there were increasing instances of active participation by the Americans from that time onward."[6]

"Approximately 40 U.S. military personnel of all ranks and services were permitted to have their families accompany them to Vietnam on a two-year tour of duty. The remaining 700 odd members of the advisory group served for one year, the maxim time which the Department of the Army thought reasonable for an unaccompanied tour in an unhealthful and distant country. Their actual tour as advisors seldom exceeded 11 months. Military Assistance Advisory Group personnel in Saigon lived in comfortable bachelor officer or enlisted quarters with good common messing facilities, and the entertainment and recreational opportunities of Saigon close at hand. In the field advisors occupied more primitive, although seldom uncomfortable, quarters, with few diversions except for a nightly movie in the mess hall. They also suffered fewer distractions and demands for paperwork than their counterparts in the capital."[7]

"As a general rule the lower the rank of the advisor and the farther from the capital, the closer his relationship to his unit and his counterpart. The regimental, division, or corps advisor spent a good deal of his time consulting with his Vietnamese opposites and observing and inspecting the components of his unit, either alone on tour or with the unit commander. Nor did the field advisor completely escape the demands of paperwork. His immediate supervisor and advisory group headquarters required frequent periodic reports, and the U.S. advisor often had to aid his counterpart in dealing with Vietnamese paperwork. The advisor was also likely to spend a good deal of his time traveling, since the area assigned to his division or regiment was frequently large and its subordinate units scattered. One Officer spent almost all his time in travel. He was the deputy chief of the Military Assistant Advisory Group, a major general, who regularly visited and inspected all units in the field to keep General Williams informed and to ensure at least some uniformity in the training effort."[8]

"A good deal of the advisor's effort was directed toward persuading his counterpart of the wisdom of certain policies, procedures, or assignments he wished the Vietnamese to adopt. Aided by an informal system of communications, advisors at the higher echelon would pass the word on what measures had been applied, and in some cases aid the lower ranking advisor by persuading

MAAG Advisor Col. Nat Ward, Chief Of Staff, MAAG Saigon, 1958-1960, stands next to unknown LT. Col. in Saigon, 1959. He is presenting him with a Certificate of Achievement. Both men wear the abbreviated summer khaki uniform, with the black web belt and gold roller buckle. (Nat Ward Collection)

Col. Nat Ward wearing the Class B summer khaki uniform, which is the same as the Class A, but without the black four-in-hand cotton/mohair necktie. MAAG Advisory Compound, Saigon, 1959. (Nat Ward Collection)

their Vietnamese opposites to direct their subordinates to take desired action. Yet, in an Army where orders of superiors were frequently ignored or evaded, such methods were far from completely effective."[9]

"The 11 month tours tended to militate against advisor effectiveness. As one officer recalled, 'A Vietnamese division commander would have an advisor for eleven months, and then he would get a new one. The new one would have to start from the zero point again. (The Vietnamese commander) had heard everything before, and he knew that the advisor didn't understand the language and that the advisor couldn't be everywhere all the time to see what was going on… He knew all about how to handle advisors.'"[10]

"Because the short tour separated an officer from his family, advisors often considered it a hardship and a nuisance. 'One consequence of the short tour was that we had a number of people who merely wanted to get through,' observed one former senior advisor. 'They'd spend three months getting orientation into their jobs, and the last three months getting ready to leave.'"[11]

The following are directly out of the Country and Study and Station Report Republic of Vietnam by the Military Assistance Institute, Department of Defense, 1960.

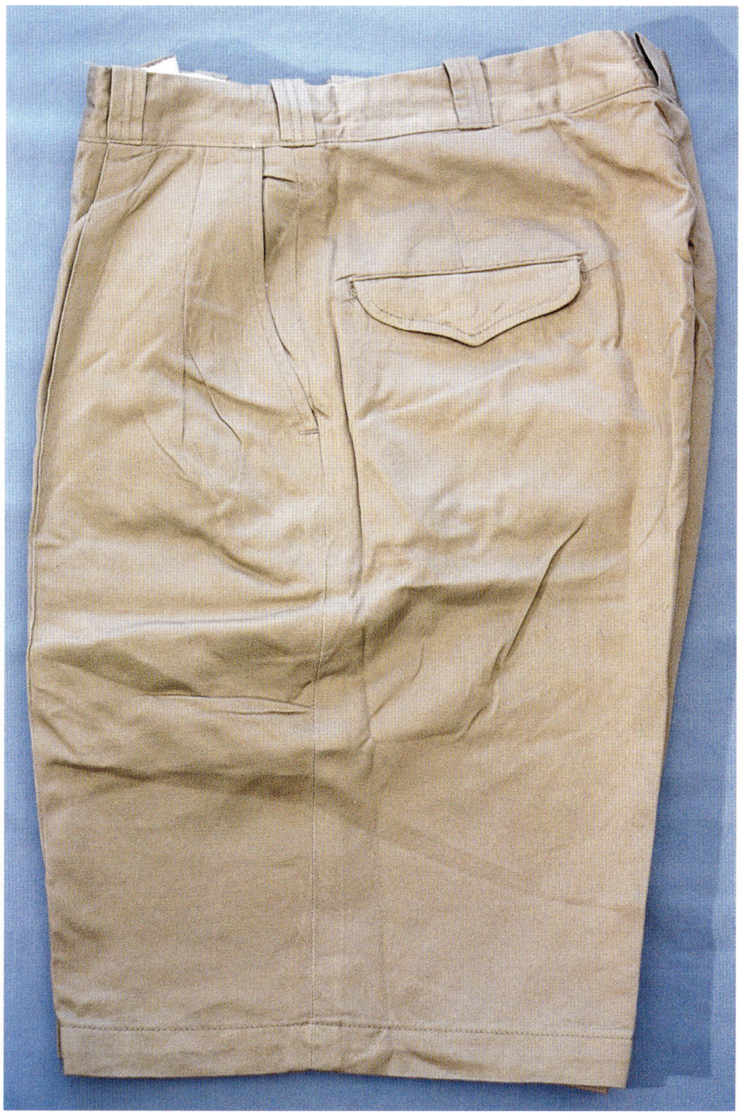

Close-up showing the abbreviated cotton tan shorts.

Distinctive items of uniform and uniform accoutrements:

Various Arms and Services are designed by the color of the beret worn with the uniform. The type of unit is designated by the scope of the shoulder patch worn on the uniform. Known colors are the following:

Arm or service	**Color of Beret**
Navy	Ultramarine Blue
Marines	green
Air Force	dark blue
Rangers	Maroon
Airborne	Scarlet (red)
Instructors (schools and training centers)	black
Infantry and Service Troops	Olive Brown (Tan)

The shoulder boards are the same for all officers regardless of service. They are black with gold insignia for company grade and silver for field grade.

Cadets where red shoulder boards with gold insignia.

Bandsmen where black shoulder boards with a gold harp.

Fatigue uniforms are worn only by Marines and airborne troops who wear the camouflage fatigues, while all other troops wear the regular olive green hearing bone twill, fatigue uniform.

Patches: The shoulder sleeve patches are made in various shapes which indicate the type unit headquarters to which the individual is assigned. Colors vary with the unit, but carry no special significance as far as unit identity.

Unit	**Shape of shoulder patch**
Joint General staff	oval
Capital military zone	oval
Corps	round
Divisions	shield
Service Troops	square

Use of distinctive piping and colors for various units has been discontinued. (as of 1960)

Vietnamese Navy

The Vietnamese Navy (VNN) came into being in 1952. In 1955 it was officially activated as an independent arm of the national defense establishment. The mission of training the Vietnamese Navy was taken over from the French by MAAG Vietnam in 1957. (See Chapter 5)

Decorations

Medals and other outward signs of recognition for military or civilian achievement were a crucial part of Sino-Vietnamese civilization. Their influence was further heightened by the cultural impact of the French, who also prize awards. The Vietnamese soldiers take great pride in their medals, and the lower ranks will wear the whole decoration rather than just the ribbon. With the re- establishment as an independent state in 1949, the new government proceeded to create new, distinctive decorations for its armed forces. Four awards were established: the National Order, the Medal of Military Merit, the Cross of Valor, and the Medal of Honor and Merit. There was also a special medal for civilian meritorious acts.

MAAG

MAAG Vietnam Memorandum No. 341-1, 14 May 1962, governs use of Air Post Office (APO) and should be consulted before sending gifts and other items subject to customs regulations. All personnel should advise correspondents to address mail as indicated below:

Headquarters MACV:
Master Sgt. Albert Thomas
J3 division
Headquarters MACV
A PO 143, San Francisco, California

Headquarters, MAAG
Major Thomas Jones 0-11111
Eng. Br. Log Div, Army section
Headquarters MAAG
APO 143 San Francisco, California

Advisory Team
Major John R. Jones
Advisory Team#29
APO 38
San Francisco, California

Advisory team Number	appropriate APO address
1,2,3,4,5,6,7	APO 137, San Francisco, California.
21, 22, 23,	APO 95, San Francisco, California.
27, 28, 29	APO 38, San Francisco, California
34, 35, 34, 35,	APO 40, San Francisco, California
36, 37, 38, 39	APO 40, San Francisco, California
33, 40, 41, 42	APO 15, San Francisco, California
33, 51, 52, 51,	APO 15, San Francisco, California
52, 53, 54, 55,	APO 27, San Francisco, California
56, 57, 58, 59	APO 15, San Francisco, California
60,61,62,63,64,	APO 15, San Francisco, California
70......................	APO 27, San Francisco, California
74, 75, 76,77, 78, 79	APO 143, San Francisco, California
80, 81, 82, 83, 84, 85, 86, 87	APO 143, San Francisco, California
90, 91, 92, 93, 94, 95, 96	APO 143, San Francisco, California

The following are Public and Social facilities of MAAG responsibilities:

Ban Me Thuot, Bien Hoa, Can Tho, Vung Tau, Dalat, Danang, Duc My, Hue, Kontum, Nha Trang, Pleiku, Quang Tri, Quang Trung, Qui Nhon, Saigon, Song Mao, and Thu Duc.

Ban Me Thout:

Uniform: Khaki, long or short. The fatigue uniform, with boots, is required for field training exercise. Washable suits and sports clothing are desirable for off duty wear. Dress for evening meals is informal.

Recommended Supplies: 3 sets of short khaki w/long socks, 2 overseas caps, 2 pair of low-quarter shoes, 3 sets of long Khaki, 3 sets of fatigues, 2 fatigue caps, 2 pair of boots, one rain coat, pair of tennis shoes and shorts, several washable sports shirts and trousers, several weeks' supply of postage stamps, shoe polish, brass polish, and cigarettes.

Bien Hoa:

Uniform: Cotton khaki, abbreviated or semi-abbreviated, sun helmet, service cap or garrison cap worn year round as normal duty uniform. Fatigues worn as required. No requirement for winter uniform. Senior Advisors occasionally prescribe dress uniform for certain important functions. The individual officer should have one dark business suit for wear to various VN gatherings. Dress normally is informal.

Recommended Personal items: Sunglasses, maps, camera, shotgun, hunting rifle, film, recreational equipment, liquor, and snacks as desired by individuals. Beer and soft drinks locally available at low prices. (Each BOQ has a refrigerator) Radios, record player, and tape recorder are available for the individuals' use.

Close-up showing the pocket configuration of tan abbreviated shorts and tan/OD green belt with gold roller buckle.

Close up showing the 1955 pattern khaki shade 1 short sleeve shirt with the box pleat pockets and pointed flaps. First produced in 1955.

Can Tho:

Uniform: The cotton khaki uniform is prescribed for wear during duty hours. Wearing of civilian clothes during off duty is optional. On some (although rare) occasions, such as formal receptions, the wearing of the dress white uniform is required. White uniforms can be made to measure by a local tailor for approximately 12 dollars. This includes a white cover for the service cap, but does not include brass buttons.

Recommended personal items: A month's supply of toilet articles, including a safety razor, is recommended. Soft drinks and beer are available on a reimbursable basis. Candy, peanuts, and alcoholic beverages for personal consumption should be brought if desired.

Dalat:

Uniform: Due to the cool climate the uniform for the advisors to the military academy is long sleeve khaki shirt and trousers with tie. Short sleeve khaki shirt with long trousers is an optional uniform. The tropical worsted uniform with blouse, winter green uniform, or white uniform are sometimes required for ceremonies.

Recommended personal items: During off hours the men will be required to wear sports clothes, to include short and long sleeve sports shirts, slacks, etc. A warm jacket and sweaters are needed during the evenings. A sports jacket and slacks, wool business suit, and one lightweight suit are recommended.

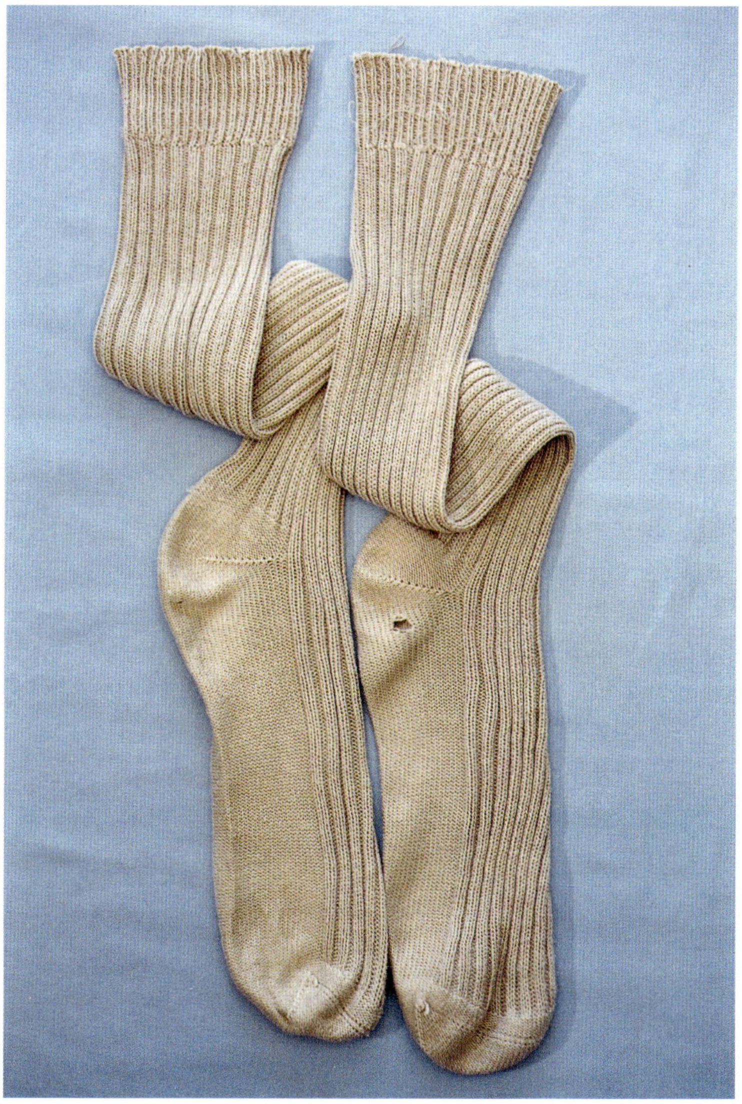

Close-up showing the Sand Shade 115 knee length socks, which were introduced with the other components of the abbreviated uniform on 1 July, 1956.

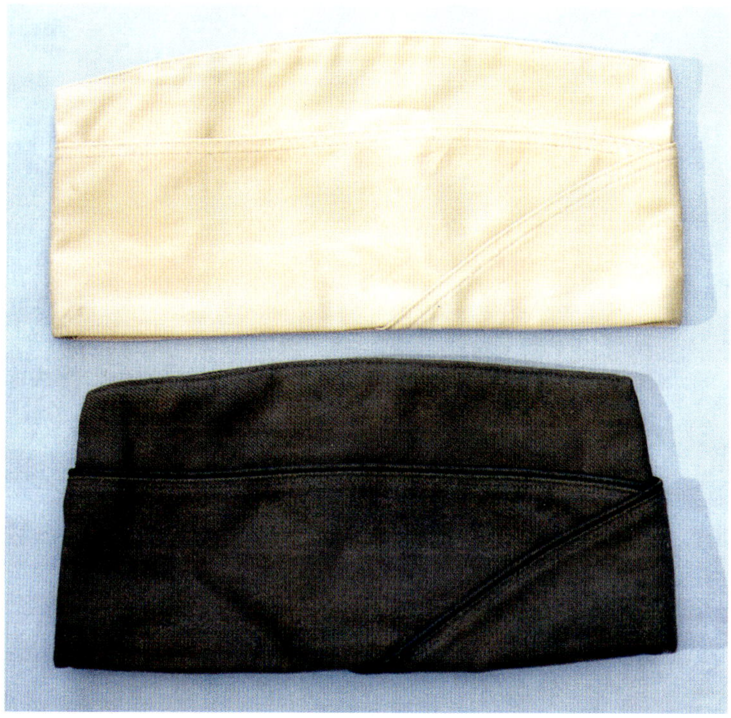

Close-up of the M1950 khaki garrison cap (top) and the green garrison cap (bottom), which began to phase out the Khaki cap beginning in 1957 (for officers), and by 1958 everyone was expected to wear the green cap. Khaki was no longer permitted for wear.

Cap, Garrison, Khaki
M-1950
7 1/8
Miller Bros. Hat Co. , Inc.
Dated 6-11-52
QM-28103-01-16368
Mil-C-1912A
#73-C-18070-50
Patt. Date 12-5-50
N.Y.Q.M.P.A.

Close up of tag for the Cap, Garrison, khaki M-1950. Note the cap is dated 6/11/52.

Danang:

Uniform: MAAG memorandum No. 14, 1960, prescribes the uniform to be worn. Advisor assigned to the I Corps and second infantry division, plus some of the advisors from the second to military region, will need several sets of khaki and fatigue uniforms and one pair of combat boots. Ponchos, raincoats, rubber boots, and field jackets are available for issue by the supply officer. It is customary for the advisor to wear the VN unit insignia on the left pocket of the shirt.

Recommended personal items: Personnel reporting for duty at this station should arrive with sufficient toilet and other personal items to provide for their needs for at least a three-week period. Although not all encompassing, the following list will serve as a reminder of the type of items which are included in this category: Hand soap, shaving cream, toothbrush, deodorants, talcum powder, shoe polish, shoe laces, stationary, film and photo supplies, lighter flints, sewing kit, alcoholic beverages, bleach, soap powder, starch, razor blades, tooth paste, shampoo, after shave lotion, heat rash powder, shoe brush, cigarettes or tobacco, stamps, lighter fluid, polishing cloth, soft drinks, and coat hangers.

In addition to the foregoing, and to those items required to conform with this station's uniform regulations, it is recommended that personnel reporting for permanent duty at this station provide themselves with at least four changes of field uniform (HBT) shirts and trousers.

Required supplies: The following items of organizational clothing and equipment will be issued at this station to personnel reporting for permanent duty: jacket, shell, field, lightweight poncho, pistol belt, Canteen, Canteen cover, can, meat, CRS, case, First aid packet, Fork, knife, spoon, field mess, steel helmet, helmet liner, rubber boots, pocket, ammo magazine, flashlight, cargo field pack, and carrying bags with sling.

Duc My:

Uniform: Duty uniform is the fatigues, or summer service, as prescribed by the senior advisor. The abbreviated uniform may be worn. Advisors usually attend ARVN official ceremonies in the service uniform (long sleeve khaki shirt with tie) and service cap. Infrequently the white uniform is worn. After duty hours appropriate civilian clothing may be worn.

Recommended personal items: Newly assigned personnel reporting to Duc My should bring a two-week supply of toilet articles, writing material, stamps, cigarettes, and film. Additional supplies can be obtained weekly from the Navy exchange and Commissary.

Hue:

Uniform: The duty uniform is the fatigue uniform. Organizational clothing and equipment will be issued upon arrival. The khaki uniform with long sleeve shirt and long trousers is occasionally the required uniform for Vietnamese functions. Advisors wear the MAAG insignia on the fatigue uniform. It is essential to have a white uniform for certain social functions all advisors attend.

Recommended personal items: Personnel reporting for duty at this station should arrive with sufficient toiletries and other personal items. Although not all-inclusive, this serves as a reminder of the items which should be brought initially: hand soap, shading cream, toothbrush, deodorant, talcum powder, shoe polish, shoe laces, stationery, film and photo supplies, flints, sewing kits, hair tonic, razor blades, toothpaste, shampoo, after shave lotion, heat rash powder, shoe brush, cigarettes or tobacco, stamps, lighter fluid, polishing cloth, and coat hangers.

CAP, GARRISON,
WOOL, SERGE,
AG. 44 EM.

MFG. CO. INC.
DATED 19 DEC. 1957
DA-36-243
QM (CTM) 1548
MIL-C-14012A (QMC)
PATTERN DATE
22, JULY 1957
MC&TSA100% WOOL

Close-up of tag for the Cap, Garrison, Wool, Serge, AG 44EM. Note this cap is dated 1957.

Close up of Low Quarters, Shoe, Man's.

Close-up showing the Army's Khaki cotton Class A summer shirt if worn with a necktie, and Class B if no necktie is worn. Note the shirt features the old-style E-7 rank.

Close-up showing the pocket configuration for the Army's khaki cotton Class A trousers.

Kontum:

Uniform: With it is recommended that personnel assigned to the detachment have a minimum of the following: three sets of fatigues, two long sleeve khaki shirts, three short sleeve khaki shirts, three long khaki trousers, three shorts khaki, two pair boots, one white dress uniform or tropical worsted dress uniform with shade 33 service cap, and one rain coat.

UNIFORMS:

"A short sleeve khaki uniform shirt was produced according to specifications on 19 December 1955. The shirt was introduced as a component of the abbreviated khaki cotton uniform on 1 July 1956 and became a mandatory item of khaki attire commencing 1 July 1958. The shirt was made of 8.2-ounce cotton twill in khaki shade No. 1. It was designed with an open style collar without a neck button. The shirt has shoulder loops, two locks pleat pockets with buttoned pointed flaps, and a square bottom with side bands at the hem. Later issue was modified to simplify production and conformed to Army style conventional shirt patterns. The box pleat and pointed flap were eliminated from the pocket designed effective 1 July 1966. The long sleeve khaki uniform shirt was limited before wearer with the uniform. On the 7th 1966, in anticipation of this change, DA directed soldiers to truncate long sleeve shirts. This was accomplished by cutting the sleeves off evenly, turning under the raw edge, and stitching a new hem. The hem was set high enough to prevent the sleeve from wrangling in the band of the elbow."[12]

"The conventional khaki uniform trousers were also made of 8.2 ounce cotton twill in khaki shade 1. The design was standard, without pleats or cuffs, and contained two side pockets, two slit-type hip pockets, a watch pocket that was later eliminated, and a button or slide fastened fly closure. The 1st pattern trousers had a self-waist-band with narrow belt loops. Later versions had a set-on waistband with wide belt loops.

The khaki cotton shorts, made from the same material as conventional trousers, were produced under specification adopted on 28 October 1955. They were first issued on 1 July 1956 as a component of the abbreviated khaki uniform, and were only worn in this fashion. The prescribed length of the khaki shorts was one to two inches above the kneecap. The shorts were initially produced with sewn in creases, but this feature hampered cleaning and was discarded in May 1957."[13]

"The original headgear with the Army khaki uniform was a service cap in matching fabric. Officers wore a cap with a soft roll grommet and a removable cover. A band of basket weave braid in khaki shade 1 encircled the cap, which had a leather visor (Slanted 45 degrees) in russet shade 90 and a matching 5/8-inch wide chin strap and head strap. The issue service cap for enlisted men, introduced in a new form during 1946, consisted of one cap frame, along with covers for each seasonal uniform. The removable cotton cover could be laundered and stretched."[14]

"The M1950 garrison cap was of standard curtain design without a gusset along the center seam. It was worn in travel status, and when prescribed by unit commanders. During times of mobilization, the service cap was not issued because of cost and convenience, and the garrison cap served instead as the principal headgear. The garrison cap was made of matching khaki shade 1 cotton twill fabric and had cord-edge braid along its upturned curtain. The braid was in the branch color of the assigned unit for enlisted men, and was gold for generals, gold intermixed with black for commissioned officers, and silver intermixed with black for warrant officers."[15]

"Headgear color change from khaki in accordance with the Army transition to the army green uniform accessories. The army

TROUSERS, MEN'S, COTTON UNIFORM TWILL KHAKI 2101, TYPE II, CLASS 1 (FINISHED BOTTOMS)
DSA-1-8990
100% COTTON

Close up of the printed tag inside the: Trousers, Men's, Cotton Uniform Twill Khaki 2101, Type II, Class 1 (Finished Bottoms).

Col. Nat Ward III stands in front of an airplane at an unknown airstrip in 1959. He wears the Army's Khaki Cotton Class B summer uniform with the black web belt, black low quarter shoes, and the new Army Green Service Cap with visor. (Nat Ward collection)

green service cap and garrison cap became mandatory for officers on 1 May 1957, and officers were required to wear them after that time. Enlisted men wore the caps as they became available for issue, which was accomplished by 1 February 1958."[16]

"The army green uniform coat was designed with the same single breasted, four pocket style already in place with the white uniform coat of 1938 and the khaki semi-dress coat adopted 1942. The coat had semi peaked lapels and shoulder loops, and buttoned down the front with four large regulation buttons. The outside pocket flap was pointed in the center in on each corner.

Coat ornamentation for commissioned officers and warrant officers included a three-quarter inch braid of black mohair around each cuff, except that the braid for general officers was 1 1/2 inches wide. This was a major change in the width of the braid that had existed since before WWI for designated commissioned officers. The width was changed in accordance with Army postwar policy that reduced the equivalent trimming on women's uniforms."[17]

"The Army green uniform coat in Army green shade 44 was a winter wait coat made of wool serge, wool elastic, or wool gabardine. The first material used was 14 to 18 ounce wool serge, but 16-ounce wool elastic was adopted for the coat in 1959. Finally, on 11 January 1961, an eleven-ounce wool gabardine was announced for coat production, along with another wool serge in 12-ounce weight."[18]

"Several different shirts could be worn with the Army green uniform. The tan shade 46 shirt, made of 3.5 to 4 ounce cotton poplin, was adopted by QMCTC 3-55 and specifically designed to harmonize with Army Green shade 44 used in the Army green uniform. The tan coloration was slightly lighter than the khaki shade 1 cotton poplin shirt, which was substituted until the newer tan shirt became available in late 1957. Both shirts had a stand up collar, no shoulder loops, a left breast pocket, and one button cuffs."[19]

"The Army green uniform had matching headgear, usually made of the same fabric as the coat and trousers. For officers the army green service cap was also available in fur felt in Army Green shade 244, such as the better 'Luxembourg' commercial brand. The enlisted service cap was issued in wool serge Army Green shade 44. The service cap had a hard roll grommet, no head strap, and a plain black leather visor with a shell cordovan finish. The service caps for field grade officers (majors and above) were furnished with a black fabric covered visor having two arcs of gold bullion oak leaves. Officers' service caps had a light brown pigskin or sheepskin 5/8 inch wide chin strap covered with two vellum gold wire lace, and enlisted caps had a black leather chin strap."[20]

"The quality appearance of officer model caps, however, was frequently eroded by tarnishing of the gold color embroidery and strap coverings. This condition was often aggravated by contact with sulfurous materials through improper handling and storage, and by the use of plastic rain covers. The effect of this corrosion process was a dirty green patina, actually favored by some 'crusty' field grade officers."[21]

"The garrison cap was made of Wool Serge in Army Green shade 44. Optional fabrics and different weights were also authorized for officer uniforms. The issue Garrison cap enlisted men was designed with stiffened interfacing to keep its shape, which prevented the soldiers from peaking their caps by crushing the center portion of the crown—a traditional practice that had been disallowed in 1956. The Garrison cap had a new, straight curtain design that diminished the curvature of earlier versions. The cord edge was gold for generals, gold intermixed with black for commissioned officers, and silver intermixed with black for warrant officers. The traditional system of branch colors on enlisted garrison caps was eliminated, and the cord edge braid was prescribed in non-contrasting Army Green Shade 45."[22]

"The waist belt was the Olive Drab Shade No. 3 web belt with GCM buckle. On 29 August 1957 officers were directed to switch to a black web belt or optional woven elastic belt. Effective 13 June 1958, enlisted men were authorized to wear the black web waist belt, except in formation. On 1 October 1958 this belt became part of the initial clothing issue for soldiers, and on 15 April 1959 it officially replaced the olive drab belt."[23]

"The trousers were worn with the black military low quarter shoes or bloused into black combat service boots. Commercial black leather shoes in plain toe, blucher oxford, chukka boot, or

Vietnamese Corps anniversary party in the fall 1958. Col. Ward and the Vietnamese General had a good working relationship together. Col. Ward and the other American advisors all wear the Class A uniform with the black tie, MAAG shoulder sleeve insignia, and black web belt with brass roller buckle. Note that Col. Ward is the only one wearing his ribbons and decorations on his uniform. (Nat Ward collection)

Col. Nat Ward on the left, looking very much like General MacArthur, wears the Army's Green Service Cap with visor, Class A khaki cotton summer uniform with blacks shoes, and a riding crop is in his right hand, along with a Vietnamese made rain jacket. The Vietnamese Army Officer standing next to him, wears a set of the ARVN Olive Drab uniform with an American Ridgeway cap and tan web belt. In the background are horses of the BMT pack train, along with the corral II Corps. 1958-59. (Nat Ward collection)

1958, II Corps. 4th military region, South Vietnamese pack train in the Central Highlands.

similar design were allowed. Commercial synthetic corfram material with leather finish was permitted on optional footwear commencing in 1969."[24]

"The original khaki summer uniform included a long sleeve shirt and trousers. On 1 July 1956 DA created the 'abbreviated' cotton uniform as a second, or alternate category. The new category included a khaki cotton short sleeve shirt, khaki cotton shorts, and sand shade 115 knee length cotton socks. The Army had experimented with this form of dress for use in the tropics during WWII, but was forced to abandon the idea of a short sleeve shirt and knee length trousers during the war because of lack of malaria protection. The Army reintroduced the shorts when civilian 'Bermuda shorts' were in vogue."[25]

"The abbreviated khaki uniform was popular at first but restricted to warmer geographical regions. Off-duty wear was restricted to areas where man's shorts were considered appropriate during hot weather. The uniform variant remained an available summer category, but it quickly fell out of favor after civilian fashions abandoned the Bermuda shorts look. On 5 April 1962 DA regulations declared that the uniform could no longer be worn in formation, and the Army ceased issue of the shorts."26

"The final category of the Army khaki uniform was created by combining the khaki short sleeve shirt with the conventional trousers. Commanders had the discretion of allowing wear of the short sleeve khaki shirt with the trousers by 27 June 1957. With the introduction of the lightweight army green uniform, the wear of the long-sleeve shirt with necktie as the Class A summer uniform ceased. On 23 September 1966 DA regulations announced that the short sleeve shirt was the only shirt permitted with the trousers. This action created the need for short sleeve khaki shirts, and as a result, the sleeve was truncated by DA directive."[27]

"Army shade 115 knee length socks were introduced as a component of the abbreviated khaki cotton uniform on 1 July 1956. The socks, which were worn only with the khaki cotton shorts, became a mandatory khaki item on 1 July 1958. The seamless socks were made of combed two-ply cotton knitting and nylon reinforcing yard. They were worn about 1 inch below the bottom of the kneecap and a turnover of about two and a half or three inches at the top."[28]

"Footwear consisted of the russet low quarter bluchard Oxford pattern shoe or combat service boots into 11 September 1956,

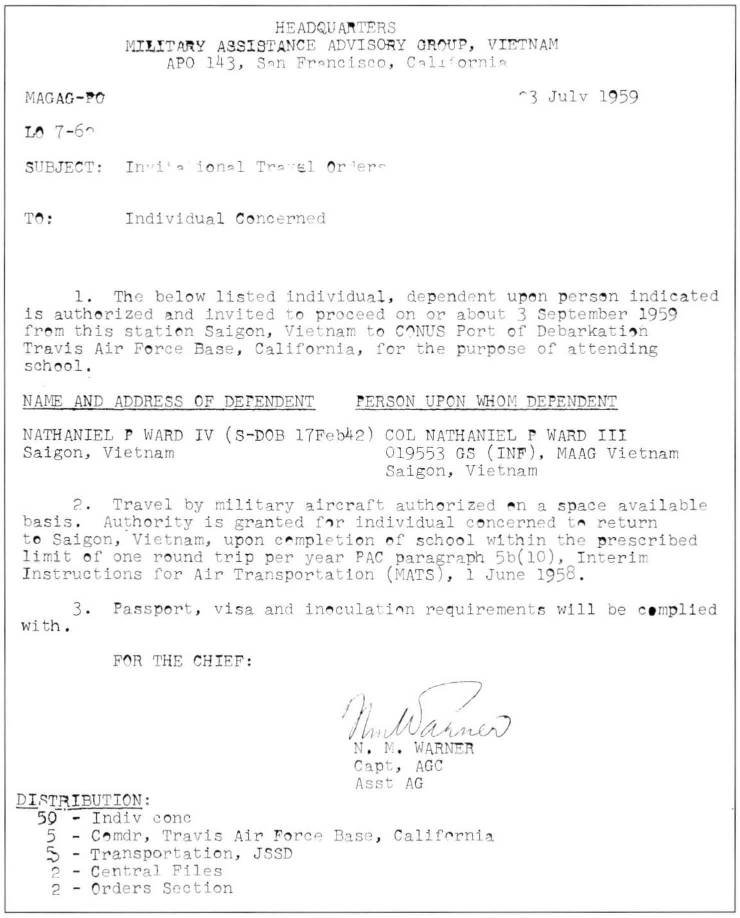

HEADQUARTERS
MILITARY ASSISTANCE ADVISORY GROUP, VIETNAM
APO 143, San Francisco, California

MAGAG-PO 23 July 1959

LO 7-60

SUBJECT: Invitational Travel Orders

TO: Individual Concerned

1. The below listed individual, dependent upon person indicated is authorized and invited to proceed on or about 3 September 1959 from this station Saigon, Vietnam to CONUS Port of Debarkation Travis Air Force Base, California, for the purpose of attending school.

NAME AND ADDRESS OF DEPENDENT — PERSON UPON WHOM DEPENDENT

NATHANIEL P WARD IV (S-DOB 17Feb42), Saigon, Vietnam — COL NATHANIEL P WARD III, 019553 GS (INF), MAAG Vietnam, Saigon, Vietnam

2. Travel by military aircraft authorized on a space available basis. Authority is granted for individual concerned to return to Saigon, Vietnam, upon completion of school within the prescribed limit of one round trip per year PAC paragraph 5b(10), Interim Instructions for Air Transportation (MATS), 1 June 1958.

3. Passport, visa and inoculation requirements will be complied with.

FOR THE CHIEF:

N. M. WARNER
Capt, AGC
Asst AG

DISTRIBUTION:
50 - Indiv conc
5 - Comdr, Travis Air Force Base, California
5 - Transportation, JSSD
2 - Central Files
2 - Orders Section

International Travel Orders and Logging, for Col. Nat Ward, Chief of Staff, MAAG, July 1959. (Nat Ward collection)

one black she was or combat boot became required. Can socks with low quarter shoes were also changed to black at the same time."[29]

U.S. ADVISOR, MAAG, MR II, 1959

"Before World War II the Army sharply distinguished between field service attire and necessary work apparel. At the time of national mobilization, the soldiers' traditional work clothing, or fatigues, consisted of blue denim one piece and two piece suits. The HBT Olive Drab Herring Bone twill outfits during the army's mechanization and wartime expansion replaced this. Throughout the global campaigning of World War II, modified HBT attire functioned as a 'battle dress' uniform for both work and combat purposes. By 1949, the field uniform consisted of a variety of HBT jackets and trouser combinations and one piece HBT suits for mechanics. All of these outfits continued to be worn well into the Cold War era. Several patterns were adopted through the years, but these coexisted within Army supply channels, and issue was commonly intermixed. Thus, the various components were worn without any styling conformity by officers and enlisted alike."[30]

"The HBT jacket, made of cotton herringbone twill in olive drab shade 7 was designed with a single breast button closure and a convertible collar. The wartime jacket contained upper cargo pockets or center pleat pockets, while the simplified 1946 pattern had patch pockets. Worn by this re-created MAAG team member is the Herringbone twill fatigues that are the WWII style with the large K Ration pocket and 13 star buttons. The shirt with a single breasted button closure that was a metal tack button, usually referred to as the "burst of Glory," or the 13 star buttons in japanned black were gradually replaced by flat phenol formaldehyde (plastic) sew on buttons in olive drab shade 6."31

"The pocket was a large cargo style, which was intended to hold the K-Rations, but many soldiers found this method of carrying their meals to be uncomfortable. These large pockets would soon be replaced by the 1945 version, which featured smaller patch pockets. The matching trousers had two large cargo pockets, one on each hip, which featured a center pleat. The HBT were designed to fit 'loosely' for field comfort and movement,' and the ample cargo pockets could hold a variety of items, in addition to the boxes of K-Rations."[32]

Above his left pocket is the "U.S. Army distinctive tape, first authorized 27 Oct. 1953. The woven label has golden yellow block

U.S. Army advisor, MAAG Vietnam, II Corps, 1959.

Full figure of the MAAG advisor with emphasis on his right side. Note the russet brown holster for his .45 pistol, and WWII double buckle boots, which were still in use in 1959.

Full figure view of American advisor, MAAG, wearing the herring bone twill uniform.

Upper torso shot provides a good look at his herring bone uniform, pin-on C.I.B., and one piece MAAG Vietnam patch.

lettering on a black background. A nametape was added by local unit directives above the right pocket and contained the soldier's last name, first initial. Early name tapes were often in branch colors, but most were made of white engineer tape with black lettering."[33] Above the U.S. ARMY strip is a metal pin-back C.I.B. "This Major wears the single gold leaf of his rank and the yellow crossed rifles of his branch (Infantry) on right and left collar points."[34] Advisors from 1955 to 1964 wore the blue and white MAAG VIETNAM shoulder sleeve insignia until control was switched over to the control of MACV. On his right shoulder he wears the SSI of the 8th Army, showing a combat tour with them during the Korean War from 1950-1953.

"Field gear at this time for advisors was largely a mixture of World War Two/Korean War era items, along with the recently introduced M1956 web gear, depend-

Close-up of the MAAG Vietnam shoulder sleeve insignia. The patch was taken from the thirteen star blue field from above the head of the Eagle on the Great Seal of the United States. This can also be found on the 13 star metal buttons on the various U.S. uniforms.

Close-up of advisor's waist showing the M1944 OD pistol belt, with the Korean War era magazine pouch and WWII era russet brown holster on his right side. On his left side is a WWII Marine Corps jungle first aid kit, a woodsman's pal survival knife/machete, and an aluminum canteen with cover.

Left: *Upper torso shot of advisors, with emphasis on his right side provides a good look at his early name tape and shoulder sleeve insignia on the right sleeve, showing a combat tour.*

ing on the availability."[35] Advisors were usually armed with a variety of surplus U.S. weapons from World War II, such as Tommy guns, M3 "Grease guns," and used here the M1A1 folding stock carbine, which was very popular with both the Americans and Vietnamese alike. "The choice of weapon would usually dictate the type of web gear adopted."[36] Web gear worn here consists of an M1944 pistol belt with a Korean War era pouch to hold four 30 round banana magazines for his carbine. Worn on the left hip is the Canteen, Corrosion, Resistant Steel, and is seen here in the felt lined Cover, Canteen, Dismounted, M1910."[37] A

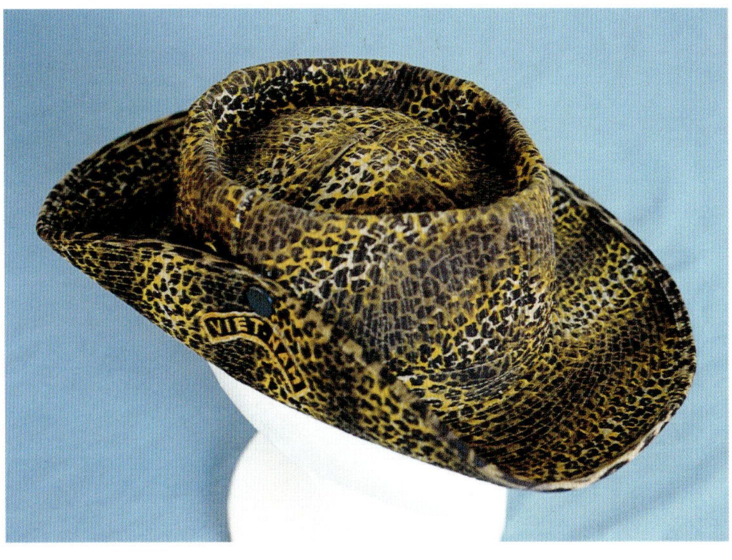

Close-up of advisor's odd leopard pattern cowboy hat. Note the black and gold Vietnam arc on the side with the button.

Additional close-up of the leopard pattern cowboy hat. Note the three eyeholes on the side.

Close-up of advisor's WWII brown leather double buckle boots.

Additional close-up showing the rear of the boot, along with the green cotton laces. These were still in service, even though the Army had done away with brown leather footwear.

russet brown M1936 holster for his .45 cal. M1911A1 auto pistol, a WWII Jungle first aid kit, and oddly enough a 'Woodsman Pal' utility knife complete the gear.

Headgear is an example of the "cowboy" hat favored by both the American personnel and Vietnamese alike. This example is in the very rare "leopard" pattern, named because of its likeness to the animal. The hat features 3 vents/eyeholes on each side, a snap located on the left which allows the wearer to secure it to the crown, giving an Aussi look, and an additional silk woven VIETNAM arc is sewn slightly above the snap on the left side.

The advisor wears World War II Brown Leather Double Buckle boots. Although the Army had "officially" done away with all brown leather boots and chose black leather instead in 1958, these boots would be worn in Vietnam as late as 1965 in some cases.

Change of Command Ceremony in Saigon for the Vietnamese Navy, 1959. (Nat Ward collection)

2

U.S. Army (MACV) Advisors 1963-1971

"With over ten years of experience by 1965, the American advisory effort in South Vietnam was well-established. The focus of the effort, however, had changed during the past five years from advising a peacetime army to advising one at war. Obviously, the exigencies of assisting a war time army differed markedly from those involving one at peace. But the basic role of the advisor was unchanged. Advice remained advice, and, although accountable for both the advice they gave and the receptivity it received, advisors remained advisors, with no legal command responsibilities. Any changes in their relationships with their South Vietnamese counterparts involved far-reaching political, rather than military, decisions."[38]

"The U.S. Military Assistance Command, Vietnam (MACV), was the supreme American military headquarters in South Vietnam. Although technically a sub-theatre command, it combined elements of both a country Military Assistance Advisory Group (MAAG) and an independent joint operational command. MACV not only supervised an extensive field and staff advisory effort, but also controlled several combat commands, such as the U.S. Army 5th Special Forces Group (Airborne) and the U.S. Air Force 2nd Air Division. In addition, MACV had a growing number of political responsibilities and unusual state-of affairs brought about by the deep involvement of the South Vietnamese officer corps in local and national politics, and by the politico-military nature of the war itself. These combined political, advisory, and operational responsibilities presented unique challenges to the MACV headquarters and the officers who led it."[39]

"Prior to the deployment of American combat units, a U.S. Army advisory group assisted each of the four South Vietnamese corps headquarters. These groups supervised not only the local division advisory teams, which in turn controlled the regimental, battalion, province, and district advisory detachments, but also the five area logistics command advisory teams until their subsequent transfer to MACV J-4 in December 1965. The major mission of each field detachment was to advise the Vietnamese commander and staff of the unit, to which was assigned all aspects of military operations and to coordinate all direct American assistants for that unit. The advisors were also the eyes and ears of

Group of South Vietnamese officers checking construction on a road near Trung Lap, Vietnam, Circa 1962-63. The Vietnamese officer at far right wears a set of tigerstripe with his rank on the left pocket flap, while the officer in the middle wears the traditional "Duck Hunter" pattern. The remaining four officers wear U.S. style OG 107 cotton sateen fatigues with a variety of headgear. The two men in camouflage both wear the Vietnam infantry brown/tan beret, while the man in the center wears an ARVN style cap. The two officers on the left wear U.S. M1 helmets with the WWII netting, and a Ridgeway cap with his rank pinned to the front.

Captain Mink (left) and Lieutenant Xuon (right) somewhere near Cu Chi. Both men wear tiger stripe fatigues, while Captain Mink wears a white T-shirt and tan Vietnamese Infantry beret, and Lieutenant Xuan wears a French style cowboy hat and an M1944 pistol belt. Both men carry M1 carbines.

Close-up of a typical ARVN style service cap, OD green in heavy cotton.

Headquarters of the U.S. Army Vietnam, MACV Advisors School, Long Bihn, Vietnam. (Don Driftmeier Collection)

Close-up of printed tag inside the Vietnamese made headgear. Note the 06 62 is the date it was manufactured.

MACV headquarters, enabling it to keep tabs on almost every South Vietnamese unit and command. MACV itself provided most of the administrative support for the scattered detachments, but funneled most supplies and equipment to them through the South Vietnamese chain of command. For food, building materials, and other items, individual detachments had to rely on the local marketplace and their own ingenuity. A Corps senior advisor, usually a senior colonel, headed each corps advisory group and reported directly to General Westmoreland. However, with the arrival of large U.S. combat units in South Vietnam and the creation of U.S. corps level headquarters-I and II Field Forces, Vietnam, and III Marine Amphibious Force in the I, II, and III Corps Tactical Zones in 1965 and 1966, the authority of the corps advisory group commander greatly diminished. As each U.S. corps level headquarters was established, General Westmoreland placed the local advisory group under its direction. The group commanders became deputy senior advisor, and the new corps level commander, usually a Lt. Gen., became the corps' senior advisor. In this way the advisory groups and their subordinate teams became part of the local U.S. tactical chain of command. This included not only the division, regimental, and battalion advisory detachments, but also the province and district teams. As a result, the bulk of the field advisors lost their direct operational link with MACV headquarters. Only in the IV Corps, where no U.S. ground combat units were quartered, did the local corps advisory group remain directly under MACV."[40]

"Other advisors continued to enjoy a certain amount of independence. In addition to the four corps advisory groups, MACV retained separate advisory detachments for the follwing: the Capital Military District; the Territorial Forces; the Airborne, Ranger, Armor, and Artillery Commands; the school and training center; and the railway security force."[41] "MACV assigned each Team a number, but the size, the function, and composition varied greatly.

Full figure view of American Infantry Advisor, Team 10, 1st ARVN Division, 1964.

Rear view of American advisor shows the PRC-10 radio.

For example: Advisory Team 99 consisted of the South Vietnamese 25th division advisory team and all subordinate regimental and battalion teams; Advisory team 90 included the Tay Ninh Province team and all subordinate district teams; Advisory Team 101 assisted the South Vietnamese 40th Engineer Base Depot in Saigon; and an unnumbered team of a few officers advised the South Vietnamese National Military Academy at Da Lat. Few kept or retired any personal records."[42]

"By 1965 MACV had stabilized the size of the basic field advisory team. Battalion and district detachments consisted of five men each, and province teams were now 20 strong; regimental teams remained small, a mere 3; and division and corps teams averaged 52 and 143, respectively, with most members providing administrative support to a smaller number of key staff advisors. Although personal shortages sometimes reduced the actual strength of the advisory teams by as much as 50 percent, their authorized strength remained fixed at these levels for the next four years."[43]

"MACV expected the advisor to befriend and influence his counterpart within the context of the existing South Vietnamese politico-military system. It was not his task to solve, or even address the deeper political and social problems that beset South Vietnam and its armed forces."[44]

"Even prior to 1965, the role of the American tactical advisor had expanded far beyond that of military advice to their counterparts. As early as 1962 they had begun to serve as combat air support coordinators, directing American fixed and rotary wing support in the field. This responsibility had given them greater

Close-up of typical private purchase knife manufactured in Solingen, Germany. These knives could be bought at local hardware stores back home or in the Army PX. They almost always came with a natural stag handle and dark brown leather sheath.

control over battlefield firepower and transport, and, as a byproduct, more influence over Vietnamese plans and operations. The growth of this influence, however, was concomitant with the advisors receiving tactical radios (AN/PRC-10s) that enabled them to communicate directly with supporting aircraft. The continued expansion of combat operations and the growing number American support units available severely overextended the MAP equipped advisory communications system, and by 1965 it was barely able to support the tempo of combat activity. Nevertheless, the provision of air support was one of the chief responsibilities of the tactical advisors."[45]

"The field advisors also had a third major task: gauging the performance of the South Vietnamese armed forces and thereby providing MACV with a rough idea of how the war was progressing. Because of the small size of the field detachments, their internal staff and administrative capabilities were weak. But throughout 1964 face-to-face talks with superiors became less frequent, and written reporting requirements of the advisory teams more demanding. As the war grew in size and complexity, so did the paperwork demands on individual advisors, and some were preparing as many as 40 reports every month. Two became major bellwethers of the war effort: the Senior Advisor Monthly Evaluation (SAME) report, which forced advisors to rate each South Vietnamese unit being advised; and the province monthly evaluation report, prepared jointly by the MACV sector senior advisor and the USAID province representative. Higher military commands all the way up to Washington scrutinized both, and an advisor's career often seemed to depend on the content and acceptability of these reports."[46]

"The American troop buildup in 1965 forced other changes in the lives of the field advisors. Almost overnight, the corps, division, and province advisory teams began furnishing liaison officers to U.S. units operating in the territorial jurisdiction of the Vietnamese commands that they advised. In addition, they found it necessary to establish 24-hour operations centers similar to those in U.S. troop units so that tactical information could be exchanged at a moment's notice. Advisors at all levels also found themselves forced to spend more time addressing such matters as law-enforcement, moral and recreational issues, post exchanges, living facilities, and press and community relationships in response to policies and procedures established by American commanders. As the size of the U.S. ground commitment increased, these requirements increased proportionately, and, without additional personnel, the field detachments had to perform these tasks at the expense of their normal advisory duties."[47]

"Another important aspect of the advisory system was the length of the assignments, or tour, in South Vietnam. The Joint Chiefs of Staff originally had authorized a two-year tour for advisors with dependents and, because of the 'primitive living conditions, dietary effect...., isolation involved, adverse climatic conditions, and health problems,' an unaccompanied one-year tour for those assigned outside of Saigon was authorized. Early in 1965, with the increased tempo of combat activity, the President of the United States ordered all military dependents home, leaving U.S. military personnel with a uniform one-year tour. In the field, corps senior advisors further shortened specific assignments by having subordinate advisors serve in hardship posts, especially in battalions and remote districts, for more than six months, after which the individual rotated to the new position, normally on a higher level staff."[48]

An American S.E. Asian scholar, Gerald Cannon Hickey, found, based on research and interviews of some 400+ advisors, that some soldiers were just not cut out to advise a foreign military, no matter how good or professional they were as soldiers. "Hickey suggested many remedies: screening advisor candidates for aptitude in foreign languages, ability to work with foreign nationals, and availability for extended overseas assignments; doubling the MATA course length to 12 weeks; and adding 800 to 1000 hours of Vietnamese language instruction to the pre-tour preparation. Hickey also felt that the practice of using the advisory network as the prime source of statistical information on the South Vietnamese armed forces made little sense and only overworked advisors. Such data ought to come directly from the South Vietnamese Joint General Staff. Finally, he recommended: a nine-month field tour for advisors; sometimes longer, overlapping tours; a training program in South Vietnam to supplement the MATA course; centralized historical files on advised units and their commanders; and greater communication between advisors at all levels."[49]

"While the field advisors had legitimate grievances, most felt that their presence was necessary and beneficial. Differences in personalities, backgrounds, and missions also meant that each individual advisor had his own unique style and experience. One young advisor firmly believed that he and his colleagues at the battalion level performed the bulk of the real advisory work. The battalion advisor, he felt, is in a key position....He doesn't look to anybody for help and, because of this, he doesn't have to write reports and tie himself down with administrative business. He is forced into the Vietnamese Army; if he is effective, he eats with them and sleeps with them. This is where you get the advising done. In the young advisor's eyes, the corps, division, and province advisory teams were only burdens to the Vietnamese, who spent too much time just trying to placate the Americans—trying to either keep them informed or come up with some pet project that some American was trying to jam down their throats—rather than fighting the day-to-day war. Such sentiments were common. In the field, at the 'nuts and bolts' tactical level, the American military advisor somehow seemed more in his element; here he left the bureaucratic concentrations behind and could personally see, touch, hear, smell, and taste what was going on, things that were only words and statistics at the echelons above him. As he planned operations with his counterpart, helped him train his troops, and called American air support when needed, he became an integral part of the unit he was technically only advising, and in turn, the Vietnamese battalion commanders regarded the advisor as their personal support element, providing radio communications, armed helicopters, aerial resupply and medical evacuation, and intelligence. Many even considered any lengthy combat operation without their American advisor as almost suicidal."[50]

MACV ADVISOR, ADVISORY TEAM 58, CHUONG THIEN, VIETNAM, 1963-1964

"By the end of 1961 the U.S. Military Advisory Assistance Group (MAAG) had more than tripled to nearly 3,200 soldiers, and the numbers continued to escalate. Within six months there would be some eight thousand American Military personnel in Vietnam, and by December 1962, more than 11,300. Along with this growing stream of men came a flood of equipment. The U.S. provided the South Vietnamese armed forces with small arms and machine guns, radios, and other communication equipment, transport aircraft, naval patrol boats, and armored personnel carries. The greatly enhanced U.S. Military role in South Vietnam required a more substantial command operation, and in February the MAAG mission was superseded by a new Military Assistance Command Vietnam (MACV). Paul D. Harkins, the fifty-seven year old three star General selected to take over the new command, had begun his Army career 40 years earlier in the Cavalry troop of the Massachusetts National Guard. During World War II he served as assistant chief of staff to General George S. Patton."[51]

"It would be Harkins' job to bring together thousands of American advisors with their Vietnamese counterparts and oversee the build up of American combat support units. He was also commander of the U.S. Military Assistance Command, Thailand, with responsibilities for SEATO planning and the defense of Thailand and Laos."[52] "Harkins and his staff had their counterparts in the Vietnamese Joint General Staff (JGS), and Americans filled advisory staff positions at each of the corps and division headquarters. At the regimental level, a senior advisor with a staff of several assistants worked with the ARVN regiment commander in planning operations and was available for tactical advice once operations were under way. Like the regiment advisors, the battalion's three man U.S. advisory team—two officers and an NCO—participated in the planning and execution of operations. But they also regularly accompanied their units into the field for periods of six to eight weeks, becoming the primary reporters on the scene for the U.S. chain of command."[53]

"MACV made a major effort in 1962 to extend its advisory mission to provincial administration and defense. Sector advisory teams—composed of a Lieutenant Colonel or Major, an Intelligence sergeant, a light weapons specialist advisor, an admin-

Full figure view of advisor with emphasis on his left side provides a good look at his Tommy gun and MACV shoulder sleeve insignia.

Full figure view of advisor with emphasis on his right side. This provides a good look at his Tommy gun, brown leather russet holster, and 7th infantry patch showing a combat tour with them during the Korean War.

istration specialist, a communications specialist, and a medical specialist—were dispatched to assist 36 of the 44 provincial chiefs. With a broader range of responsibilities than other American advisors in Vietnam, the sector teams not only provided advice and training in the employment of local militia units, but also played a major role in the implementation of local pacification plans and civic action programs."[54]

Although the U.S. Army had officially adopted the new Jungle fatigues in 1963, the demand for the uniform greatly outweighed the supply. Many at the top were able to acquire several sets, while the field operators like this Staff Sergeant were at the bottom of the pecking order and would have to wait. Some were able to obtain the new lightweight uniform through trading captured Chi Com weapons or flags, while others resorted to buying them on the black market, or buying them from the supply officers directly. Either way, this Staff Sergeant is wearing the 1st model OG 107 cotton sateens. On the left shoulder is the newly adopted MACV SSI. The insignia was designed to reflect its advisory role: the red background indicated Communist aggression, and the yellow sword breaking through the Great Wall of China represented America's willingness to stop Communism. Below the SSI are U.S. produced E-6 chevrons. Worn above his left pocket is the black and gold U.S. Army tape, a pair of basic airborne wings, and a C.I.B. Worn above his right pocket is a single white name strip. On his left sleeve is the SSI of the 7th Infantry Division, which shows this NCO served with them in Korea during the war. Underneath his OG 107 shirt he wears a white cotton T-shirt.

Headgear is the 1942 pattern cap, which very much resembles the Marine Corps style utility cap, but features a pleat through the center. These hats were first produced in HBT during the 40s, and then in cotton sateen during the 50s. This one is a later style, being produced in cotton sateen. These were replaced by the patrol cap, and later the dreaded baseball cap. Although some advisors like this one continued to wear what they liked, some even had these hats locally produced in both tigerstripe and the duckhunter patterns.

Web gear is an assortment of World War II and Korean War items. A set of M1945 OD green suspenders are connected to his M1944 pistol belt, along with a single pouch which holds the shorter magazines for his Tommy gun, two M1910 metal canteens, a russet brown leather holster for his M1911 pistol, and a single MKII grenade. On his back he carries the AN/PRC-10 Radio with its own carrying harness. The PRC-10 was still common to both the Army and the USMC in 1966. It was unreliable, and not well suited for Vietnam's jungles and mountainous terrain. The PRC 10 would soon be replaced by the AN/PRC-25 in the coming months, although some units, such as the 1st Inf. Div. and the 1st Cav., arrived with the newer radios.

The PRC-10 radio would be replaced a year later in select line units by the PRC-25, but the Vietnamese and their American Advisors would have to make due with these radios until 1967-68, when there were enough to begin outfitting the ARVN with the newer radio. Once again the uncomfortable all leather boots were worn while in the field, while some high ranking "Staff Pukes" had two pairs of the new comfortable jungle boots, which were probably spit shined for when he went to the Officer Club at 18:00 hours.

Photo of Capt. Smith, MACV advisor, wearing OG 107 cotton fatigues with full color insignia to include a shoulder sleeve insignia of MACV, U.S. Army Ranger tab, a U.S. Army strip, airborne wings, and CIB. On his left collar is what appears to be the Branches Of Service (BOS) of Air Defense Artillery. He is also wearing the second model jungle boots. Circa 1964-65.

U.S. ARMY ADVISOR;
TEAM 33,
Ban Me Thout,
1965

"To the Vietnamese, the American advisors were simply 'dai dien' (representatives) of the U.S. government. In the field, most understood the advisory function of their American counterparts and treated them almost as assistant commanders. Inevitably conflict

Command Post-U.S. Army Officers man the radio command post at Bien Gia. In the background is a church, mute witness to the war that rages throughout the land. Both men wear the 1st model OG 107 cotton sateen uniforms, with full color U.S. Army and name tapes. The Capt. Wears pin-on rank, while the Major (utilizing the PRC-25) has sewn-on BOS and rank, along with two WWII era "pineapple" hand-grenades.

Full figure shot of Team 55 advisor, 1965.

Full figure shot of Team 55 advisor with emphasis on his right side.

arose, especially when Americans failed to understand or accept the political and social roles of the Vietnamese officers, or when they were too belligerent or too demanding towards their counterparts. Most Vietnamese tended to be wary of the younger advisors, who normally had little military and no combat experience, while, in some cases, the Vietnamese battalion commander might have 10 to 20 years of fighting under their belts and had seen many advisors come and go on their one-year tours. To them, the war was always there, unending. Lacking the short-range goals that might have given them a sense of accomplishment, they found it difficult to share the need for action imparted by almost all the aggressive young American officers. Nevertheless, Vietnamese commanders highly valued the American combat support that the advisors made available, and at the division and corps levels, the detailed staff work that made their more ambitious operations possible."[55]

"The generalized selection process, limited training, and relatively short tours understandably caused many problems. Hickey concluded that the success of an advisor lay in not only his professional confidence, but also in his ability to establish rapport with his counterpart. He found the closest advisor counterpart relationships at the lowest levels, and noted that advisors at the higher echelons rarely saw their opposites informally and had an ever-increasing amount of non-advisory duties. Advisors at all levels appeared to be compartmentalized, and had little knowledge of what others were doing. The one-year tour in general and the six-month tour in operational units greatly limited the effectiveness of the average advisor, for by the time he had become accustomed to his surroundings and had established a working relationship with his counterpart, his tour with the unit was nearly over. In addition, the lack of any overlap between departing advisors and their replacements, and the absence of any historical records at these levels, made it difficult for the advisory effort to achieve continuity. Hickey, realizing that other factors also came into play, contrasted the good relations between South Vietnamese Ranger and Airborne units and their advisors with the often miserable ones shared between the U.S. and South Vietnamese Special Forces. Difficulties between the two later organizations, he felt, could be traced to internal South Vietnamese political problems, Saigon's prejudicial policies towards the Montagyard (mountain people), and the underlying racial animosity between the ethnic Vietnamese and Highland people."[56]

"The Textile research conducted during the Korean War led to the development of more durable cotton material with the same weight as the herringbone twill fabric. During 1952, cotton sateen fabric was manufactured in Olive green shade 107 and used to produce the new utility jacket and trousers."[57] "The U.S. Army used the official designation 'utility' for this uniform, which was

Upper torso shot provides a good look at the advisor's MACV patch, basic LBE, and his World War II "pineapple" grenades.

Another full figure shot of the advisor, front view.

Close-up showing the cotton sateen shirt with the metal "Burst Of Glory" buttons. This provides a good look at the advisor's Expert Infantry Badge, and Crossed Rifles, along with the First Lieutenant bar. Note the two pens secured to the front of the shirt.

Close up of advisor's ball cap with single silver bar, 1st Lieutenant.

consistent with the terminology of other services. The utility, or 'fatigues,' were intended to be loose fitting, although the shirt was normally worn tucked into the trousers and the trousers bloused when wearing combat boots. Army soldiers continued to call these fatigues because of the fatiguing duty performed while wearing them."[58] The first model OG 107 sateen shirt worn here came with the metal 13 star "Burst of Glory" buttons. Trousers also featured the silver metal buttons, and were worn with a black web trouser belt with a brass roller buckle.

Above his left pocket is the "U.S. Army distinctive tape, first authorized 27 Oct. 1953. The woven label has golden yellow block lettering on a black background. A nametape was added by local unit directives above the right pocket and contained the soldier's last name, first initial...."[59] Above the U.S. ARMY strip is an E.I.B., Expert Infantryman's Badge, which was awarded after firing expertly and performing other tasks, such as "enter the net," estimate range, Land navigation, S.A.L.U.T.E report, and disassemble and do a functions check on other Infantry weapons. On the left sleeve he wears the MACV Shoulder Sleeve Insignia. On the left collar is an early embroidered cross rifles Branch of Service (BOS) in full color (yellow) on HBT material, and on his right collar an embroidered full color (silver) 1st Lt. Bar, also on HBT material. Attached to the front of his shirt are two ink pens, which are just barley visible. Underneath, he wears a white cotton T-shirt.

Army web gear worn here consists of a cotton H-harness, connected to a M1944 OD Green pistol belt, with a WWII/Korean War era twin magazine pouch for his M-1 carbine. Connected to that is a WWII era First aid dressing pouch. On the Advisor's left side he carries a WWII era USMC "Jungle First Aid kit," which contained a glass bottle of "bug juice" (insect repellant), a small pressure dressing, a lip balm, and several "adhesive strips," or "Band Aids®." Next to that is the twin cell pouch for his .45 caliber pistol magazine. A single aluminum canteen is carried in the M1952 canteen carrier. On his right hip he has secured his Colt .45 M 1911 A1 pistol in the black leather holster. Connected to either side of his H-harness are the MK II Hand grenades, often-called "Pineapple grenades." Headgear is the much hated "baseball cap," which was a visor style cap made out of polyester and rayon gabardine cloth, dyed Olive Green army shade 106. The cap was first adopted in 1962 after a long search by the Army to replace the headgear of the Korean War era. A single silver, 1st Lieutenant's bar is worn affixed to the front. Boots are the new 2nd model DMS Tropical Combat Boot with Vibram soles, which were available in this style by 1964, having done away with the early leather re-enforcement along the top and vertically up the heel.

Weapon is a WWII era M-1 carbine, evident by the lack of the bayonet lug.

U.S. ARMOR ADVISOR, VIETNAMESE ARMOR SCHOOL, THU DUC, 1965-66

The uniform worn here by this advisor to the armor school at Thu Duc is another example of the OG 107 utility uniform, which was first produced in 1952. The fatigue shirt has two upper patch pockets with rectangular flaps, each secured with a single button. The matching cotton utility trousers have two front and two rear pockets, each secured with a button flap, and side button take-up tabs on the waist. A full color MACV patch is worn on his left shoulder, while on his combat side he wears a full color 40th Infantry Division patch, representing a combat tour in Korea. Below that is a set of E-6 stripes, Staff Sgt. Worn above his white name strip is a Vietnamese made, pin back metal Tanker Qualification badge, which features a tank, two crossed spears, and a wreath. This variation is a very crude "sandcast" metal badge. It

Close-up showing the American-made crossed rifles subdued (top) and Vietnamese made Cross Rifles (bottom), which are very crudely made and feature the clutch back design.

Full figure view of U.S. advisor to the Vietnamese Armor Corps. at the Armor School in 1964-65.

Upper torso shot of armored advisor provides a good look at the black Armor Corps. beret, with the silver NCO beret badge, Staff Sgt. rank on his sleeves, and MACV patch.

Full figure shot of armor advisor with emphasis on his right side. Note the 40th Infantry Division patch showing a combat tour in Korea.

Additional full figure view of the Armor advisor.

Close-up showing the private purchase black leather tanker boots.

Additional close-up showing the leather straps and brass buckle of the tanker boots.

Typical South Vietnamese NCO in the Armor Corps. He wears an OG 107 style shirt, black Armor Corps beret, and the bevo style NCO tank beret badge. Circa 1964-65.

Close-up showing the yellow armored cavalry scarf. This is step 1 in a three part process for rolling the scarf.

This is step 2. Note the scarf has been rolled once.

This is step 3, the final roll. Now the scarf is ready to put on the wearer's neck. This gives the scarf a "puffy look." The scarf is never worn as shown in step one.

Left: *Upper torso shot with emphasis of advisor's right side provides a good look at his black armor beret with metal NCO badge, full-color U.S. Army and name tape on his first model OG 107 cotton shirt. Note the Vietnamese made Armor Qualification Badge above his name strip and the "Skillcraft" pen connected to his pocket flap.*

Close-up of tag inside the scarf.

Close-up of armor advisor's right side, showing early white name tape and 40th Division Infantry patch, along with the yellow Armor Cavalry scarf, "Skillcraft" pen, and his sand cast South Vietnamese Armor Qualification Badge.

Close-up showing the South Vietnamese Armor Qualification badge (top) and the U.S. Armored Cavalry BOS pin (bottom).

Three South Vietnamese made Armor Cavalry patches. From left to right are the 4th, which is silk woven, the Vietnamese Armor School, which features torch on the upper left corner, and the 5th. Note the difference in the color of the tanks and the circles in which the numbers are imposed.

was usually awarded to the American advisors after serving in combat with the assigned South Vietnamese Armor unit. "A rayon bib-type scarf in branch of service color could be worn, beneath the collar of the shirt,"[60] as seen worn here in the Armored Cavalry color, Yellow. Of course, this was removed before going out to the field.

LBE consists of a M1956 cotton web belt, with a single item, his black leather holster for his sidearm, a Gov. issue M1911A1 pistol. Headgear consist of the black THIET GIAP BINH armored corps beret with the silver metal NCO beret device, which is very similar to the French and West German beret devices. This variation also came in silver bullion or white cotton.

Footwear is the very unique, private purchase "Tanker Boots." The boots featured double wrap leather straps, storm welting, and reinforced stitching, and were very fashionable and sought after amongst Tankers.

Close-up of black South Vietnamese Armor beret with the silver NCO beret badge.

U.S. ARMOR ADVISOR
4th SQUADRON, 4th ARMORED CAVALRY REGIMENT; ADVISORY TEAM 1, QUANG NGAI
1965-66

"Quang Ngai—The armored personnel carriers of the ARVN 4th armored cavalry Regiment moved through the gates of their compound against fallout on a seizing capture operation 275 miles northwest of Saigon. An hour earlier the second battalion, 4th Vietnamese infantry Regiment, had seized the village of Phu Thuan. 'The 2nd will secure the town to provide protection for the left flank of the airborne battalion coming in from the North,' Major Frank Walton, regimental advisor, explained prior to the start of the operation. As the carriers moved over the twisting, fog shrouded roads, they passed elements of the 5th airborne going the other way to seize a series of villages. Capt. John H. (Jack) Frost, advisors, and 4th squadron, 4th cavalry Regiment, said the airborne has two missions today—seize four villages, and link up with the air mobile forces being flown in by Marine choppers.

Overhead, jet fighters for the Marine Corps began their prestrike attack, sloping low over suspected Viet Cong positions with machine guns blazing. For an hour the personnel carriers continued through the rice paddies, deep furrows showing where the tracks had crossed. Suddenly, an excited Vietnamese voice crackled over the radio.

'There they go,' exclaimed SFC Ben Myers, enlisted advisor to the 4th squadron, as three figures broke from the rice paddies and headed for the jungle. Instantly, the squadron's machine guns opened up on the scattering trio. 'The VC are tricky,' commented

Close-up showing the Armor advisor's camouflage cowboy hat. This is done in the early ARVN Ranger pattern and were available at most tailor shops.

Full figure view of Armor advisor, 4th Squadron, 4th Armored Cavalry Regiment, U.S. Team 1, 1965-66.

Full view with emphasis on the advisor's right side.

Capt. Frost, 'they have underground tunnels criss-crossing this whole area, and usually they get away. Maybe this time we'll be lucky.' As the APCs moved across the open area, SFC Myers commented, 'we are advising ourselves right out of a job. Six months ago they would not have taken the carries across the field, they would have let the infantry go across first.'

As the APCs came to a stop, the crew dismounted for a ground search of the area, but the elusive guerrillas had gotten away, leaving a battered hat as a reminder that they had been there. For the remainder of the morning the APCs traveled over flat, level terrain, occasionally passing through villages where the crew would stop to administer first aid to ailing villagers. 'Now we wait,' declared Capt. Frost. 'The APCs will be held in reserve in case the recon unit in the Valley gets into a scrap and needs some help.' Throughout the afternoon the men of the APCs wait in the hot sun, as the gunfire continues below. Toward evening, the sound of gunfire diminished. 'Guess the guys did not need us after all,' said Sgt. Myers, as the line of three moved toward the carriers' position. 'We'll let the infantry go through our lines and provide cover for them, then move out.' Relaxing in his room that evening, Sgt. John Fortin, enlisted regimental advisor, discussed the role of the advisors in operations. 'The term advisor means just that. We don't tell a commander to do something, it's just suggested what to do. Then it's up to him to decide whether or not to follow our advice.

When we go on operations with ARVN units, much of our time is spent coordinating U.S. artillery support, naval gunfire, and medical evacuations,' he continued. Discussing the day's operation, Sgt. Fortin said, 'we didn't do too badly. We killed 5 Viet Cong guerrillas, captured one, destroyed 300 k of rice, and 25 pairs of boots.'"[61]

"The Tropical Combat uniform, or 'Jungle Fatigues,' was to be the primary field uniform in Vietnam after 1963. This uniform was a rare example of Army clothing that combined functional practicality with comfort and an attractive design. The quick-drying coat and trousers were made of a tightly woven lightweight twill cotton fabric. The loose fitting garments promoted body and heat ventilation and moisture dissipation. The coat was designed

Upper torso shot provides a good view of the Armor advisor's Vietnamese rank, his Branch of Service (BOS), Armored Cavalry, pinned below his rank, and his pin-on Combat Infantryman's Badge. Below his CIB is his Vietnamese made U.S. ARMY tape, worn in conjunction with his name, both worn with the slant of the pocket. Note, worn above his plastic pocket hanger of the 4th ARVN Armor Corps. is a small beer can (DI) Direct Insignia of the Vietnamese Airborne Division. These were given to all the members of the 4th Armored Cav. as a good will gesture, and helped provide cohesiveness between the two units, who usually worked with each other on operations.

Four Armored Cavalry troopers wearing a wide variety of uniforms. All four men are armed with M1 carbines, while one wears a set of Vietnamese tigerstripe and the black Armor Corps. beret, two others wear the duck hunter pattern, and the third wears Olive Drab. Circa 1965.

to be worn outside the trousers, and sleeves were allowed to be rolled up. In fact, it became mandatory to roll up the sleeves above the elbows neatly in military installations, and rolled down while in the fields. The coat features two slanted chest pockets, which allowed soldiers ease in removal of items, and two lower bellow pockets. The shirt also incorporated shoulder loops, side button take up tabs, and an 'inner gas flap.' All buttons on the garment are exposed."[62] Worn along the slant above each pocket are a locally produced subdued nametape and U.S. Army strip, something very unusual for 1965. Above the U.S. Army strip, this advisor has chosen to wear his pin-on CIB instead of a cloth subdued or color variation. Worn on the center of his shirt is his Vietnamese equivalent rank, 1st Lt, or Trung Uy. Underneath, he wears his BOS, Armored Cavalry, which features a U.S. Tank and crossed Cavalry sabers. Note, the U.S. Advisor does not wear any U.S. rank at all. Pinned to his left shirt pocket flap is a mini Vietnamese Airborne D.I. pin, which was given to all the members of the armor unit by the airborne to show cohesion and camaraderie. Below that is a very crude Vietnamese made 4th Armored Corps pocket hanger. Note the advisor does not wear the MACV SSI on his left sleeve. Note, also, where the advisor has chosen to wear his wristwatch, along with the Skillcraft pen and #2 pencil in his pockets. "The trousers also featured the 'exposed' buttons to match the shirt. They featured two hip pockets, two front pockets, and two large bellow cargo leg pockets. Inside the cargo pockets were two integral tie down straps, much like the M1951 field trouser offered. Inside the left cargo pocket was a small one-button pocket that was designed to hold a small survival kit that was never satisfactorily developed or issued."[63]

Headgear, worn here, is a Vietnamese made "cowboy" hat with a large brim, in the Vietnamese Ranger pattern. Third model jungle boots are worn here, which feature the new slip in spike protectors, as well as the nylon reinforcement along the ankles.

U.S. ARMOR ADVISOR
2nd SQUADRON, 7TH ARMORED CAVALRY;
ADVISORY TEAM 1
MACV COMPOUND, PK17, MR-1
"Operation Lam Son 283"
1965-1966

Capt. John H. Frost, senior advisor to the 2nd Squadron, ARVN 7th Armored Cavalry, watched as the APCs rolled down Highway One. Capt. Frost is down to his last few hours in Vietnam. During his tour he served as both RF/PF and Armor advisor, and has been decorated by both the Vietnamese Gov. and American Gov. "During Lam Son 283, a combined infantry-armor night assault on a VC stronghold, the command carrier had just started into the village," stated Capt. Frost. "As we neared a hedge grove

Full figure view of American advisor to the Second squadron, 7th Cavalry, Team 1 member, MACV Compound During "LAM SON 283," 1966

Full figure shot provides a good look at the advisor's tigerstripe uniform, black Armor beret with the gold officer's device, which is bevo woven, and a Vietnamese made brown leather shoulder holster.

Upper torso shot provides a good look at the advisor's tiger stripe uniform with the gold officers Armor beret badge. Note the map of the area in his hand.

Close up provides a good look at the advisor's Vietnamese rank, Dai Uy, and pin-on CIB.

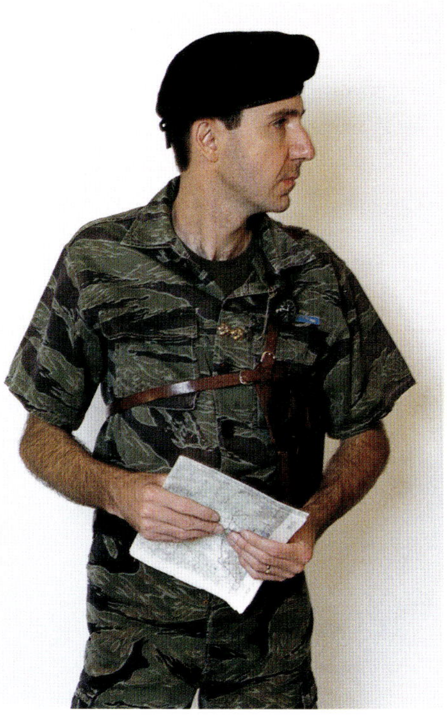

Upper torso shot shows the advisor holding a map of the area prior to operation.

An additional close-up shows advisor consulting map prior to moving out in the M113s.

Close-up provides a good look at the black web belt with gold roller buckle, his Vietnamese made brown leather holster for his .45 pistol, which also has a separate compartment for a spare magazine, his rank, Dai Uy, Timex wristwatch, which is connected to his upper buttonhole, and pin-on CIB. Note the dog tag chain, which is also secured to the upper buttonhole.

Close-up of advisor's mid section and brown Vietnamese made holster.

a VC machine gun opened up on us. The APC's driver started to back up, but was hit by a second volley of fire, and we were stuck directly in front of the gun. Sgt. Burger, Senior Enlisted Advisor, who was riding in the support vehicle further down the line of the APCs, brought his vehicle up behind Capt. Frost and exposed himself to the enemy fire as he hooked up the tow cable to pull the stuck vehicle out of the area," continued Capt. Frost.

This American advisor shows the diversity of uniforms worn by U.S. advisors in 1965-66. He is from Team 1, and was located along Highway 19. The American is part of a patrol whose job was to provide security along these lifelines that connected the government to small rural towns.

Headgear is the Vietnamese made one piece black Armor Corps beret, with the gold officer bullion Armor beret badge. Sometimes the beret badge would be worn anywhere from over the left eye to being worn over the left ear. Some American advisors took to wearing the beret badge on the left side of their cowboy hats. The advisor wears a tigerstripe short sleeve shirt and a matching set of tiger pants, which were produced in Japan. The shirt features two chest pockets with a single concealed button under the flap. Worn on the front of his shirt is the equivalent to his American rank (Dai Uy), of Captain. Note, the American does not wear his U.S. rank or his BOS on his collar points. Worn above his left pocket is a pin-on CIB. During operations, this would be removed and placed in his pocket. Connected to the top buttonhole is the advisor's dog tag chain, and dog tags are secured in his left pocket. His matching tigerstripe trousers feature two rear pockets and two thigh pockets, each with a two-button closure, along with an additional field dressing pocket, or "cigarette pocket," on the lower left leg closed with a button. All pockets are typical of Asian garments, having a single bellow gusset on one side. Around his waist is the black web belt with the gold roller buckle. The advisor's .45 caliber pistol is seen here secured in his Vietnamese made brown leather holster, which features a

Close-up, with emphasis on advisor's right side, provides a good look at his tigerstripe short sleeve shirt.

Close-up of gold officer's beret badge. (Courtesy of Jack Frost, armored advisor 1965 to 1966.)

Close-up showing the gold officer's bullion and armor Corps beret badge. (Courtesy of Jack Frost, armored advisor 1965 to '66.)

South Vietnamese armor NCO wearing U.S. style OG 107 cotton sateen uniform. Note the silver NCO beret badge is worn over his left eye. Worn over his right pocket is the Vietnamese Armor Qualification Badge, which is silk woven.

compartment for additional magazines. These were made of pigskin or water buffalo hide, and were not a high grade of leather and usually did not last too long.

U.S. ARMOR ADVISOR
4TH SQUADRON, 17th CAVALRY;
ADVISORY TEAM 1,
MACV COMPOUND HUE-,QUANG NAM
1965-1966

"The task of the ground level advisor was extraordinarily difficult. He had to be a jack-of-all-trades, not only advising, but also solving his own supply and administrative problems, often living on the local economy and training his own subordinates as best he could. Most were unfamiliar with their strange Asian allies and the type of war they were supposed to be fighting. The morale was high, but their work frustrating. With only sporadic contact with the echelons above them, they often had only their own individual resources to guide them in their unusual mission. The selection, training, and placement of the advisors themselves received relatively little attention. Prerequisites for advisor duty were generally identical to those demands for advancement in regular military service: attendance at key military schools, and successful command tours with U.S. tactical units. Colonels (corps senior advisors) were to be graduates of senior service schools, such as the U.S. Army War College, and Lieutenant Colonels (division senior advisors) and majors (regimental and province senior advisors) graduates of the U.S. Army Command and General Staff College, but neither having been passed over for advancement to the next higher rank. Prior to 1965, candidates were normally volunteers anxious to serve, the war providing assignments that offered operational experience in a combat environment. In Vietnam, MACV tried to assign newly arrived officers to units appropriate to their U.S. military backgrounds—combat arms officers to combat units, engineer officers to engineer units,

Full figure shot of American advisor to the South Vietnamese 4th Armored Cavalry, Team 1, 1965-1966.

Full figure shot with emphasis on the advisor's right side. Note the black leather holster for his .45 and the silver metal NCO Armor beret badge, worn on the black Armor beret.

Full figure shot provides a good look at this NCO assigned to the South Vietnamese 4th Armored Cavalry.

and so forth. But by 1964, with the flood of junior officers and non-commissioned officers needed to fill battalion and district advisory teams, the importance of military expertise in advisory posting at the lower levels had become irrelevant."[64]

"Preparation for advisory duty was minimal. The U.S. Army Special Warfare School at Ft. Bragg, NC, offered a short six-week military assistance training and advisory (MATA) course for advisors. The instruction stressed counterinsurgency tactics, small arms training, psychological operations, and civic action, and included some general background on Vietnam. About 1/2 of the course (120 hours) consisted of training in the Vietnamese language, and this was sometimes followed by eight to twelve weeks of more intensive instruction at the Defense Language Institute in Monterey, California. Priority for language training went to battalion, province, and district advisors who worked with their counterparts on a daily basis. But the limited capacity of Monterey, and the length of time needed to acquire working knowledge of the complex Vietnamese tonal language greatly restricted advanced linguistic training. Within South Vietnam, only district advisors received additional instruction, and others reported immediately to their assignments. All continued to rely heavily on the Vietnamese military interpreters assigned to almost every detachment."[65]

This Armor advisor wears a set of the second model "Jungle fatigues," which were identical to the first model, except the but-

Close-up showing the advisor's second model shirt with the 4th armor Corps shoulder sleeve insignia, Staff Sgt. rank, along with Vietnamese produced U.S. Army strip, U.S. full color Airborne wings, and barely visible, pin-on C.I.B. Note the BOS worn on the right pocket flap.

tons on the pockets have been covered to prevent snagging. They still featured the epaulets, side take up tabs, and a gas flap. These were produced for about a year, before the epaulets were removed, along with the side take up tabs, which makes the second model as sought after and desirable as the first with the exposed buttons. The matching second model trousers have the covered buttons and feature the button take up tabs on the waist. Worn as an SSI is a silk woven Vietnamese armor squadron patch. The patch features a white and dark blue shield with a light blue border with a light blue and white tank. Variations existed with light blue parts, as well as violet, gray, and dark blue. At the top is a dark blue disk with light blue border and a white number 4. Printed, handsewn, and later, subdued variations of this patch were worn, but stayed the same shape with the shield divided in typical Cavalry format.

Headgear is the black Vietnamese Armor Corps beret, which is one piece, and is worn in the American style with the silver NCO metal beret badge over his left side, almost over his ear. If he were to wear it Vietnamese/French style the tank would be going in reverse, thus implying the Vietnamese were "running away" from the battle, not charging forward!

Worn beneath his SSI is a set of U.S. produced E-6 chevrons. Above his left pocket is a crude Vietnamese made U.S. Army strip, a pair of full color U.S. made basic jump wings, and a pin-on C.I.B. which would usually be removed before an operation and slipped into a pocket. On his right side he wears an officer's

Close-up showing a South Vietnamese enlisted man's armor beret badge. Note the tank is not surround by a circle and is slightly longer.

South Vietnamese armor NCO wearing an M69 flak vest and a black Armor beret with the NCO device. Note the long AK-47 magazine in his M-16 rifle.

BOS armored cavalry pin in lieu of the Vietnamese qualification badge. Note the ever-useful Skill Craft pen, still used today by U.S. government agencies like the post office and still made by the blind.

LBE consists of a single M1956 cotton web pistol belt with a single twin cell pouch, capable of holding two 30 round magazines for his M-2 carbine and a black leather holster for his 1911A1 pistol. 3rd model jungle boots are worn .

Unknown Infantry advisor wearing 1st model jungle fatigues, towers over his South Vietnamese counterpart NCO.

Unknown U.S. Infantry advisor wearing the first model exposed buttons shirt, void of any patches, but with the addition of pin-on rank, sunglasses, and a dark brown Vietnamese Infantry beret. Note the beret device worn on the beret is actually the cap Badge. Above that, although not really visible, is the U.S. Branches Of Service, crossed rifles. Circa 1966-67.

Same unknown advisor wearing basic LBE and a bandoleer of 40 mm ammo around his waist for his M-79 grenade launcher, which he is removing from the jeep.

**U.S. ARMY INFANTRY ADVISOR,
5TH ARVN INFANTRY DIVISION,
7TH Infantry Regiment,
TEAM 70,
MR-3, Binh Duong Vietnam,
1967-68**

"One of the largest arms and ammunition caches of the war was discovered June first by 5th ARVN Division and 1st U.S. Infantry Division soldiers during a five-day operation six miles northwest of Saigon. The operation began with elements of the 4th Battalion, 8th Regiment of the 5th ARVN Division sweeping through a heavily jungled area with D Troop, 1st Squadron, 4th Cavalry of the 1st Division working as a blocking force. On the sweep the ARVN soldiers suddenly came across a VC platoon size unit and fired on them, killing four. Simultaneously, another 8th Regiment unit moved in from the opposite direction and killed 5 more. On one of the bodies were documents from the headquarters of the Dong Nai Battalion commander to his higher headquarters giving a complete report of his unit strength and arms they were carrying. With this information, a search and seal operation on the town of Dong An II, located about a mile from the

Upper torso shot provides a good look at the advisor's right side. Note the 9th Infantry patch showing a combat tour with them.

Full figure shot of U.S. Army advisor, Team 70, Binh Duong Province, 5th Vietnamese Infantry Division, 1967.

Full rear shot provides a good look at the advisor's early prototype 1967 2qt canteen with cover.

Full figure of Team 70 advisor with emphasis on his left side. Note the full color MACV SSI is still worn.

Upper torso shot provides a good look at his locally made E-6 stripes and CIB.

Upper torso shot provides a good look at the advisor's M-79 grenade launcher and the two bandoleers used to carry the 40 mm rounds.

division's Di An base camp, was planned with the 1st Division troops to seal the village while ARVN troops searched. Two Viet Cong were killed in the operation, and three prisoners of war (POWs) taken. One of the prisoners was a commander of the reconnaissance Company of the Thu Duc district. He told the ARVN troops that the village was a military and political headquarters for all Viet Cong units operating in the area. During the search of the village, numerous items were found, but the 8th Regiment commander, Lt. Col. Le Nguyen Vy, wasn't satisfied. He had his men cut sharp sticks and probe around the three graveyards surrounding the village. The first probe struck a box of ammunition, and the search was on in earnest. They found two 12.79 mm antiaircraft machine guns, one 82 mm mortar, two 60 mm mortars, 1 Chinese communist Browning Automatic Rifle, 2 Chi-Com light machine guns, 1 RPG-2 Rocket Propelled Grenade launcher, 2 Chinese communist AK-47s, 1 AK-50 submachine gun, 1 CKC Rifle, 32 B-40 charges, 14,000 rounds of ammunition, and many other items, such as field phones, blasting caps, rifle grenades, telephone wire, and weapon oil.

Right: *Close up provides a good view of the advisor's dark brown Vietnamese Infantry beret with the silver Enlisted Cap device being utilized as a beret device instead. Note the 550 para cord connected to the right side, which is used to secure his compass.*

Close-up of dark brown infantry beret, which is usually light tan, but the colors varied. Note the large silver enlisted cap badge used as a beret device on the beret and the black silk liner.

Spc. 4 Carl Farren assigned to HQ Company, 5th ARVN Division. He wears the third model OG 107 jungle fatigues with a full color 5th ARVN infantry patch on his right pocket, along with red and blue HQ colored tape above his name tape. On his back he carries the PRC-25 radio. Weapon is an M-1 carbine. (Carl Farren Collection)

Specialist 4 Carl Farren, during advisory training at the USAVN Advisory School, 1968. (Carl Farren Collection)

The following day, Col. Vy pressed the search, and his men found: 39 82mm rounds; 13 cans of mortar charges and fuses; 15 AR 15 magazines; 1 Chinese communist rifle; and several documents. Major Gordon Van Amburgh, senior advisor to the 8th Regiment, said 'That particular site was the corner area of the Di An, Lai Thieu, and Thu Duc districts, and each district chief thought the other was looking after the area. This is how the VC were able to operate so long without being bothered there. The documents found on the VC bodies the first day of the operation are really what tipped Charlie's hand and led us to this huge weapons cache, and allowed us to break up their headquarters."[66]

Uniform consists of the new "rip-stop" cotton poplin, which were basically cotton-poplin which incorporated a nylon weave that greatly increased the strength of the fabric. The new uniforms began to see service in late 1967-early 1968. Many combat

Close-up of early prototype 2qt Canteen with the rubberized canvas cover. These began to appear as early as 1967, and would be replaced in 1968 by the all nylon cover.

Backside of 2 quart cover provides a good look at the printed label; it reads "Cover, Water Canteen, Collapsible, DSA 100-67-1285."

Advisor Carl Farren sits atop a bunker at Advisors compound. He wears the third pattern jungle fatigues and the tan ARVN infantry beret. (Carl Farren Collection)

Advisor Carl Farren with "friend" in a small bar outside of the ARVN Division HQ in La Thieu. Note the tan ARVN Infantry beret is worn, along with his duty uniform, which is the jungle fatigues. (Carl Farren Collection)

troops would not see these until later in the war. These uniforms could be purchased on the black market of Saigon, or through supply in exchange for an AK-47 or a VC flag. Worn above the pockets are U.S. Army strips and matching name tape, and a hand sewn subdued C.I.B. Worn on the combat side is a 9th Infantry "Octofoil" SSI shoulder insignia, which represented the heraldic symbol of the 9th son and was lovingly called the "psychedelic cookie." E-6 chevrons are Vietnamese made and worn on both sleeves, along with a full color MACV SSI, which is worn on his service side.

Web gear consists of a cotton M1956 H-harness connected to a M1956 pistol belt, two universal pouches containing 40 mm rounds and personal items, and a single compass/first aid dressing pouch, which holds his compass. A single 1qt canteen with cover, and the new 3rd model 2qt. Canteen with the experimental rubberized cover, is worn on the back of his pistol belt. Additional rounds are carried in the flimsy cotton bandoleer worn across his chest. Around his neck is an OD green scarf.

Carried by this advisor is the M-79, affectionately called the "blooper" for the sound it made when the round left the chamber.

Advisor Carl Farren sits inside a helicopter. Note the tan ARVN Infantry beret. (Carl Farren Collection)

Advisor Carl Farren wearing third model jungle fatigues with subdued name and U.S. Army. Note the two-tone red and blue colors on his right pocket flap below his name; these are the Headquarter's colors. Also note he wears a full color 5th ARVN infantry patch on his right pocket. La Thieu, 1968. (Carl Farren Collection)

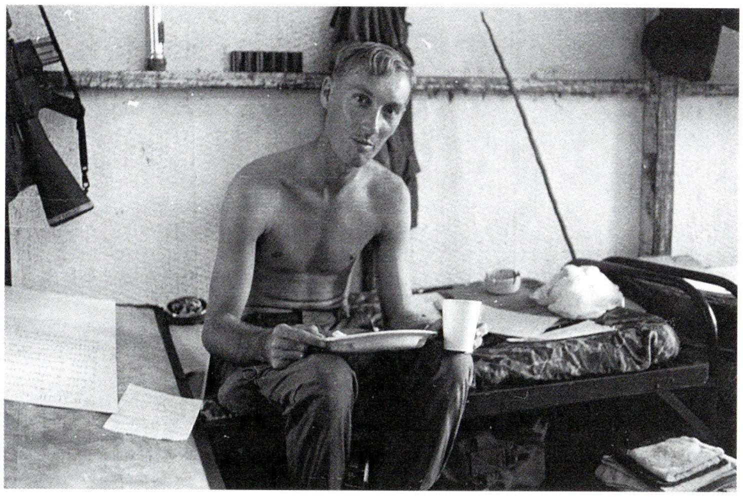

Spc 4 Carl Farren enjoys a meal on his bunk in the Advisors' compound at La Thieu, 1968. Note the early "duck hunter" poncho liner on his bunk, along with the numerous ash trays and the map board to his immediate right. Also note his M-16 with the masking tape at the end with his name on it. (Carl Farren Collection)

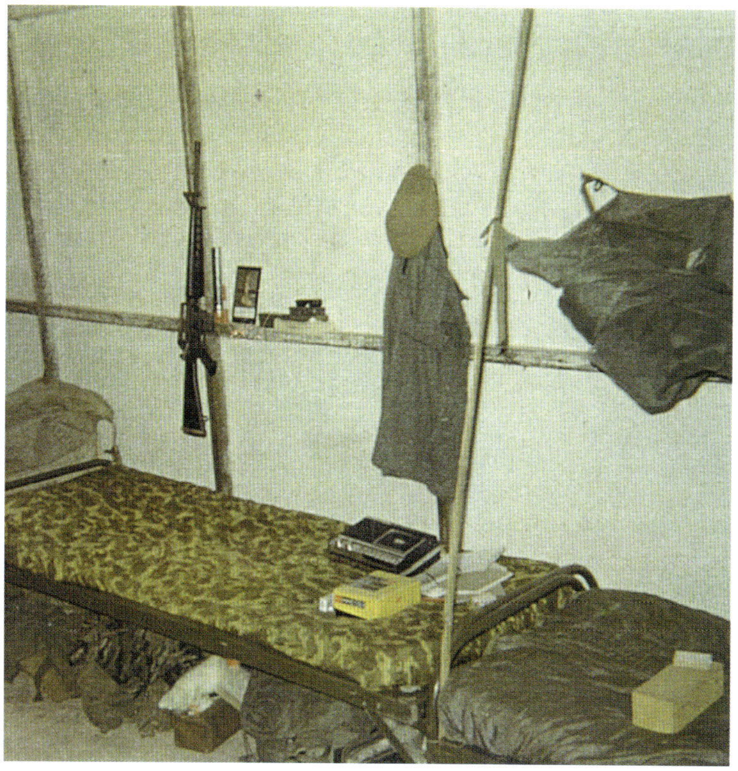

Carl Farren's bunk and personal effects at the advisors' compound in Phu Van, 1968. This is a good photo showing the early "Duck Hunter" poncho liner, along with a set of tiger stripe fatigues underneath his bunk, also. On the wall is his fatigue shirt and tan infantry beret. Personal items include a Kodak camera, tape recorder, flashlight, and numerous pens and paper. (Carl Farren Collection)

Springfield Armory had begun development of this 40 x 46 mm SR weapon in the early 1950s. A basic goal of the weapon was the replacement of the standard rifle fired grenade, and 60 mm mortar at the squad and platoon level. The M79 could cover the area between the longest range of the hand grenade (30-40 meters) and the middle range of the 60 mm mortar (300-400 meters). In the hands of a skilled grenadier, the M79 was highly accurate to 200 meters, and worked well in the jungle and in building to build-

Carl Farren wearing full uniform, including jungle fatigues, jungle boots, and his tan ARVN infantry beret, at Phu Van, April 1968. (Carl Farren Collection)

Carl Farren at the Advisors' compound La Thieu, 5th ARVN infantry division. He wears his third model jungle fatigues, which feature subdued U.S. Army and name tape, along with full color Spc. 4 rank, MACV patch, and a printed 5th ARVN Infantry Division patch. March 1968. (Carl Farren Collection)

Two members of team 70 relax after a mission around a table. Note the lounge chair and PRC 25 radio. The large square object in the right corner is the teams' barbecue. (Carl Farren Collection)

Photo of flag pole and bunker system of La Thieu, 1968. (Carl Farren Collection)

Left: *Spc. 5 Carl Farren (left) with beret, along with two senior advisors from Headquarters Company, Team 70. Behind them are the team's interpreter and another soldier. (Carl Farren Collection)*

Advisor Team 70's radio man, Spc.5 Carl Farren, and Vietnamese interpreter on a mission outside of Phu Van. Carl carries the PRC-25 radio with the headset and carries an M-16 rifle. Third model poplin jungle fatigues are worn with subdued insignia and the tan infantry beret. (Carl Farren Collection)

Left: *Vietnamese Col. of the 5th ARVN infantry division with American Lieutenant Colonel of team 70 on mission inside the Hobo Woods, 1968. (Carl Farren Collection)*

Left: *Close-up of Carl Farren's jungle fatigue shirt, which features subdued U.S. Army, his name strip "Farren," CIB, locally produced Spc. 5 rank, and MACV patch. On the right pocket is a subdued printed 5th ARVN infantry division patch, and the right is the patch of the 18th Regiment. Next to that is a Vietnamese made hammock and his tan infantry beret. (Carl Farren Collection)* Above: *Another close-up showing Coral Farren's jungle fatigue shirt. Note the blue square on the pocket—this is a platoon designator. At the top is his OD boonie hat, which was usually worn on operations, instead of the infantry beret. (Carl Farren Collection)*

ing during urban warfare. The M79's basic combat round, the M-406, was a 170-grain high-explosive fragmentation grenade at a velocity of 76 m/s to a maximum range of 400 meters. Development of the grenade-type warhead for the weapon that would become the M79 came out of the research conducted at the U.S. Army Ballistic Research Laboratories (BRL) at Aberdeen Proving Ground.

Headgear is a dark brown ARVN Infantry beret. Although the true color for the ARVN Infantry was tan, many ARVN and Americans alike wore the brown ones, also. Note the beret badge worn on the two piece beret is actually the cap device. The beret badge is actually much smaller, but for some reason the advisor wears the silver enlisted/NCO badge as a beret device.

The view from team '70's "Grosney Compound," Lam Som, March 1968. (Carl Farren Collection)

Carl Farren being awarded the Army Commendation Medal with V for Valor by the senior advisor, Col. Leach, at Lam Som, 1968. (Carl Farren Collection)

Photo of Specialist 4 Carl Farren on operations in 1968. Note he wears the Vietnamese tan infantry beret, which usually was not worn on operations, even though most of his photos show him wearing it on operations. The OD boonie hat was more appropriate, and more often than not, worn. (Carl Farren Collection)

The following items are personal effects carried by the advisor, or left with his personal effects at the advisors' compound, including the Advisory Team 70's Thanksgiving menu. Note the 5th infantry insignia on the front. This was a professionally made menu, much like the ones found in expensive restaurants or onboard U.S. ships. (Carl Farren Collection)

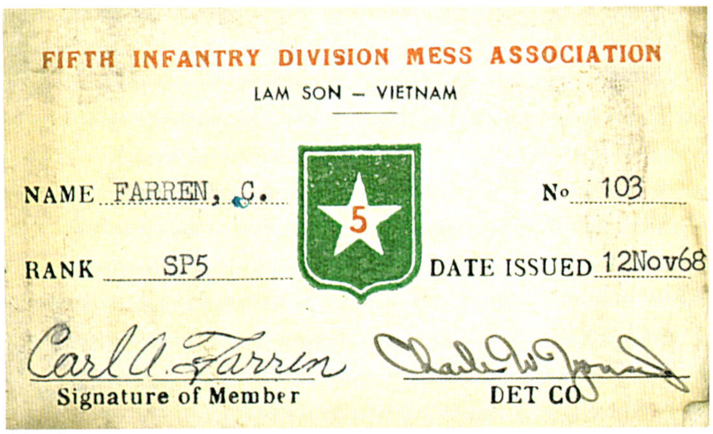

FIFTH INFANTRY DIVISION MESS ASSOCIATION
LAM SON – VIETNAM

NAME FARREN, C. No 103
RANK SP5 DATE ISSUED 12Nov68
Signature of Member DET CO

5th Infantry Division Mess Association from Lam Son, Vietnam. Carried in the wallet by the advisor. (Carl Farren Collection)

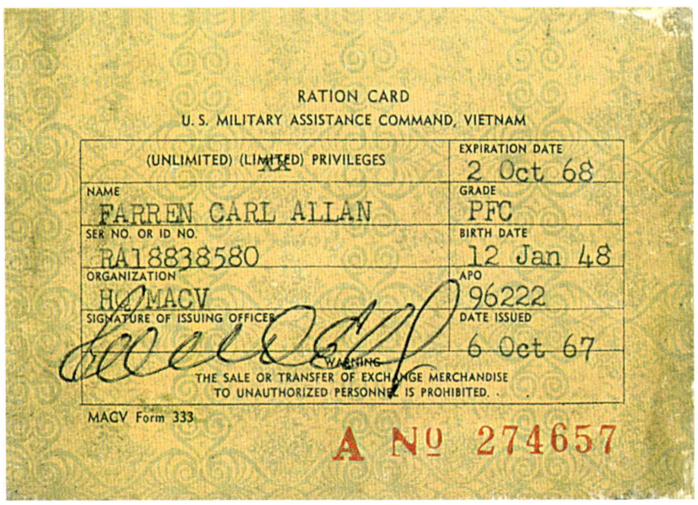

RATION CARD
U.S. MILITARY ASSISTANCE COMMAND, VIETNAM

(UNLIMITED) (LIMITED) PRIVILEGES	EXPIRATION DATE 2 Oct 68
NAME FARREN CARL ALLAN	GRADE PFC
SER NO. OR ID NO. RA18838580	BIRTH DATE 12 Jan 48
ORGANIZATION HQ MACV	APO 96222
SIGNATURE OF ISSUING OFFICER	DATE ISSUED 6 Oct 67

WARNING
THE SALE OR TRANSFER OF EXCHANGE MERCHANDISE TO UNAUTHORIZED PERSONNEL IS PROHIBITED.

MACV Form 333

A No 274657

Carl Farren's ration card. This was also carried in the advisors' wallet. (Carl Farren Collection)

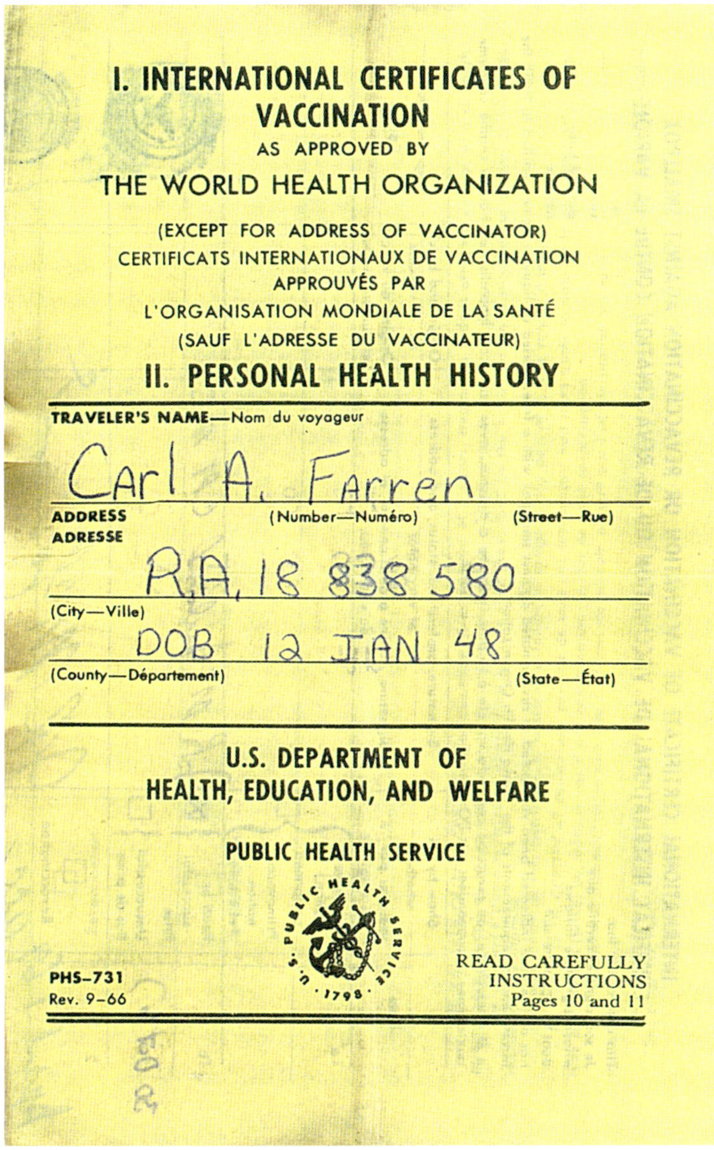
I. INTERNATIONAL CERTIFICATES OF VACCINATION
AS APPROVED BY
THE WORLD HEALTH ORGANIZATION
(EXCEPT FOR ADDRESS OF VACCINATOR)
CERTIFICATS INTERNATIONAUX DE VACCINATION
APPROUVÉS PAR
L'ORGANISATION MONDIALE DE LA SANTÉ
(SAUF L'ADRESSE DU VACCINATEUR)
II. PERSONAL HEALTH HISTORY
TRAVELER'S NAME—Nom du voyageur
Carl A. Farren
ADDRESS ADRESSE (Number—Numéro) (Street—Rue)
R.A. 18 838 580
(City—Ville)
DOB 12 JAN 48
(County—Département) (State—État)
U.S. DEPARTMENT OF HEALTH, EDUCATION, AND WELFARE
PUBLIC HEALTH SERVICE
PHS-731
Rev. 9-66
READ CAREFULLY INSTRUCTIONS Pages 10 and 11

Carl A Farren's international certificate of vaccination. This was left back at the advisors' compound during operations, but usually carried with him while at the compound. (Carl Farren Collection)

United States Military Assistance Command, Vietnam handbook. (Carl Farren Collection)

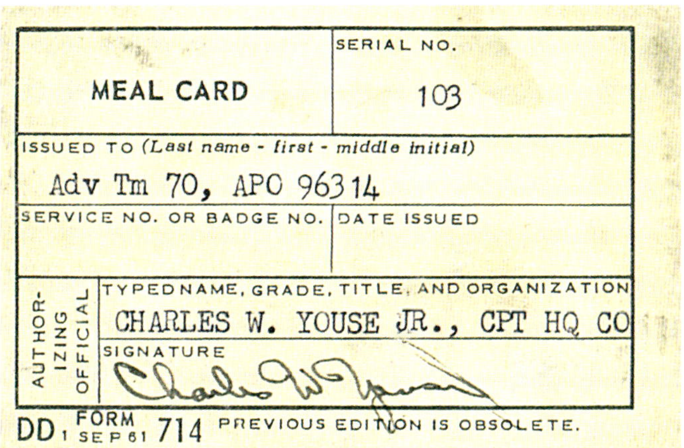
MEAL CARD
SERIAL NO. 103
ISSUED TO (Last name - first - middle initial)
Adv Tm 70, APO 96314
SERVICE NO. OR BADGE NO.
DATE ISSUED
AUTHORIZING OFFICIAL
TYPED NAME, GRADE, TITLE, AND ORGANIZATION
CHARLES W. YOUSE JR., CPT HQ CO
SIGNATURE
DD 1 SEP 61 FORM 714 PREVIOUS EDITION IS OBSOLETE.

Farren's meal card from La Thieu. (Carl Farren Collection)

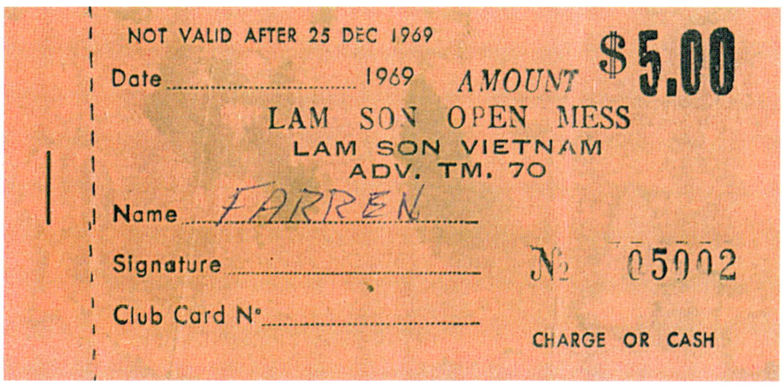
NOT VALID AFTER 25 DEC 1969
Date .... 1969 AMOUNT $5.00
LAM SON OPEN MESS
LAM SON VIETNAM
ADV. TM. 70
Name FARREN
Signature
№ 05002
Club Card N°
CHARGE OR CASH

Lam Son open mess for Advisory Team 70. This was Spc5 Farren's meal card upon arrival on his second tour. Note the price of the booklet is five dollars. (Carl Farren Collection)

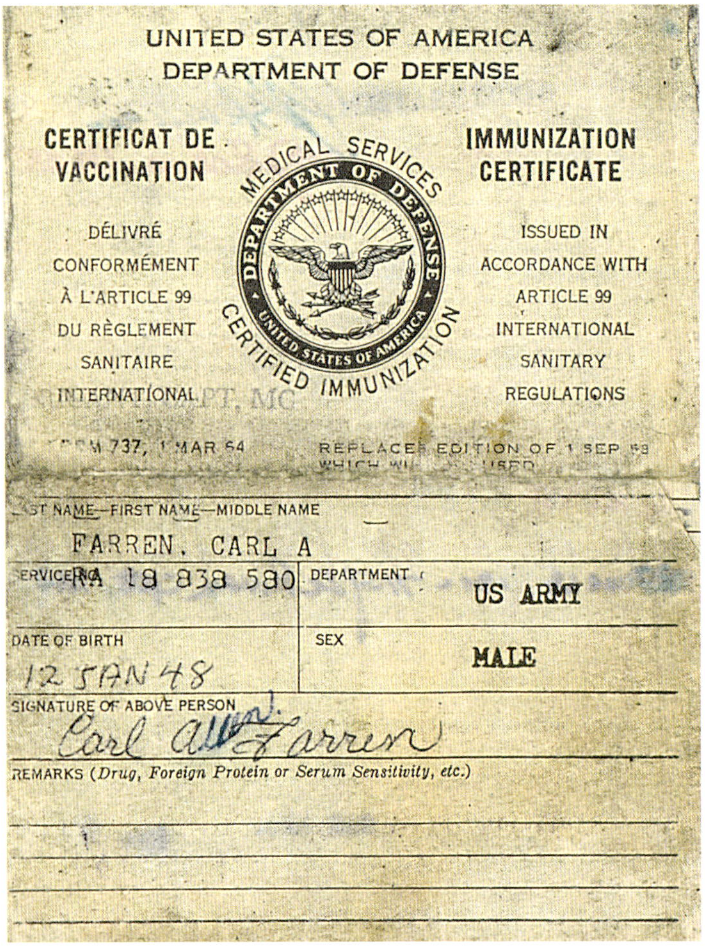

UNITED STATES OF AMERICA
DEPARTMENT OF DEFENSE

CERTIFICAT DE VACCINATION

DÉLIVRÉ CONFORMÉMENT À L'ARTICLE 99 DU RÈGLEMENT SANITAIRE INTERNATIONAL

IMMUNIZATION CERTIFICATE

ISSUED IN ACCORDANCE WITH ARTICLE 99 INTERNATIONAL SANITARY REGULATIONS

MEDICAL SERVICES — DEPARTMENT OF DEFENSE — UNITED STATES OF AMERICA — CERTIFIED IMMUNIZATION

...M 737, 1 MAR 64 REPLACES EDITION OF 1 SEP 59 WHICH WI... USED

...ST NAME—FIRST NAME—MIDDLE NAME
FARREN, CARL A

SERVICE NO. RA 18 838 580 | DEPARTMENT US ARMY

DATE OF BIRTH 12 JAN 48 | SEX MALE

SIGNATURE OF ABOVE PERSON Carl Allen Farren

REMARKS (*Drug, Foreign Protein or Serum Sensitivity, etc.*)

America Department Of Defense Immunization Certificate. Usually kept in the advisor's wallet.

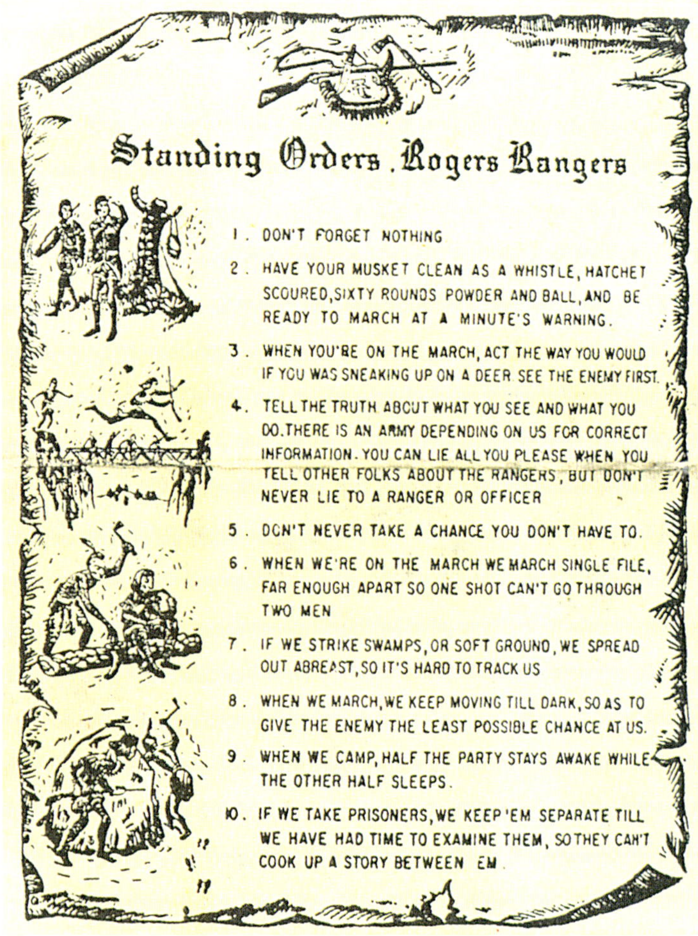

Standing Orders . Rogers Rangers

1. DON'T FORGET NOTHING
2. HAVE YOUR MUSKET CLEAN AS A WHISTLE, HATCHET SCOURED, SIXTY ROUNDS POWDER AND BALL, AND BE READY TO MARCH AT A MINUTE'S WARNING.
3. WHEN YOU'RE ON THE MARCH, ACT THE WAY YOU WOULD IF YOU WAS SNEAKING UP ON A DEER SEE THE ENEMY FIRST.
4. TELL THE TRUTH ABOUT WHAT YOU SEE AND WHAT YOU DO. THERE IS AN ARMY DEPENDING ON US FOR CORRECT INFORMATION. YOU CAN LIE ALL YOU PLEASE WHEN YOU TELL OTHER FOLKS ABOUT THE RANGERS, BUT DON'T NEVER LIE TO A RANGER OR OFFICER
5. DON'T NEVER TAKE A CHANCE YOU DON'T HAVE TO.
6. WHEN WE'RE ON THE MARCH WE MARCH SINGLE FILE, FAR ENOUGH APART SO ONE SHOT CAN'T GO THROUGH TWO MEN
7. IF WE STRIKE SWAMPS, OR SOFT GROUND, WE SPREAD OUT ABREAST, SO IT'S HARD TO TRACK US
8. WHEN WE MARCH, WE KEEP MOVING TILL DARK, SO AS TO GIVE THE ENEMY THE LEAST POSSIBLE CHANCE AT US.
9. WHEN WE CAMP, HALF THE PARTY STAYS AWAKE WHILE THE OTHER HALF SLEEPS.
10. IF WE TAKE PRISONERS, WE KEEP 'EM SEPARATE TILL WE HAVE HAD TIME TO EXAMINE THEM, SO THEY CAN'T COOK UP A STORY BETWEEN EM.

Copy of Rogers' Rangers' standing orders. This was also carried by the advisor on operations. (Carl Farren Collection)

Autographed picture of the lovely Raquel Welch. Spec 5 Carl Farren accidentally bumped into her after a USO show, and was able to get this autograph. Always carried with him! (Carl Farren Collection)

Headquarters
5th Infantry Division, 32d Division Tactical Area
Advisory Team 70, APO 96314

MAGTN-ISSA 12 January 1968

PFC Carl A. Farren, RA18838580
5th Infantry Division Advisory Detachment
Advisory Team 70
APO 96314

Dear Private Farren:

On the occasion of your birthday, I want to extend my best wishes for this day and many happy returns throughout the year.

I take this opportunity to thank you for your efforts in accomplishing our team's mission and wish you continued success in all future endeavors.

Sincerely yours,

ROBERT D. SONSTELIE
Colonel, Infantry
Senior Advisor

Headquarters
5th Infantry Division, 32d Division Tactical Area
Advisory Team 70, APO 96314

MAGTN-LSSA 18 September 1968

SUBJECT: Letter of Congratulation

SP/5 Carl A Farren
HQ, 8th Regiment
5th Inf Div Det
Adv Tm 70
APO 96314

1. It is with pride that I and Advisory Team 70 congratulate you as you move one more step up the ladder of achievement. I feel confident that you will approach the added responsibilities commensurate with your advancement with the same zeal and courage you have displayed in the past.

2. Keep up the good work!

JAMES H LEACH
Colonel, Armor
Senior Advisor

Copy of letter of congratulations to Sp-5 Carl Farren upon his promotion to E-5 from Col. James Leach. (Carl Farren Collection)

Left: *Copy of Birthday Greetings from Col. Sonstelie, Senior Advisor of team 70. (Carl Farren Collection)*

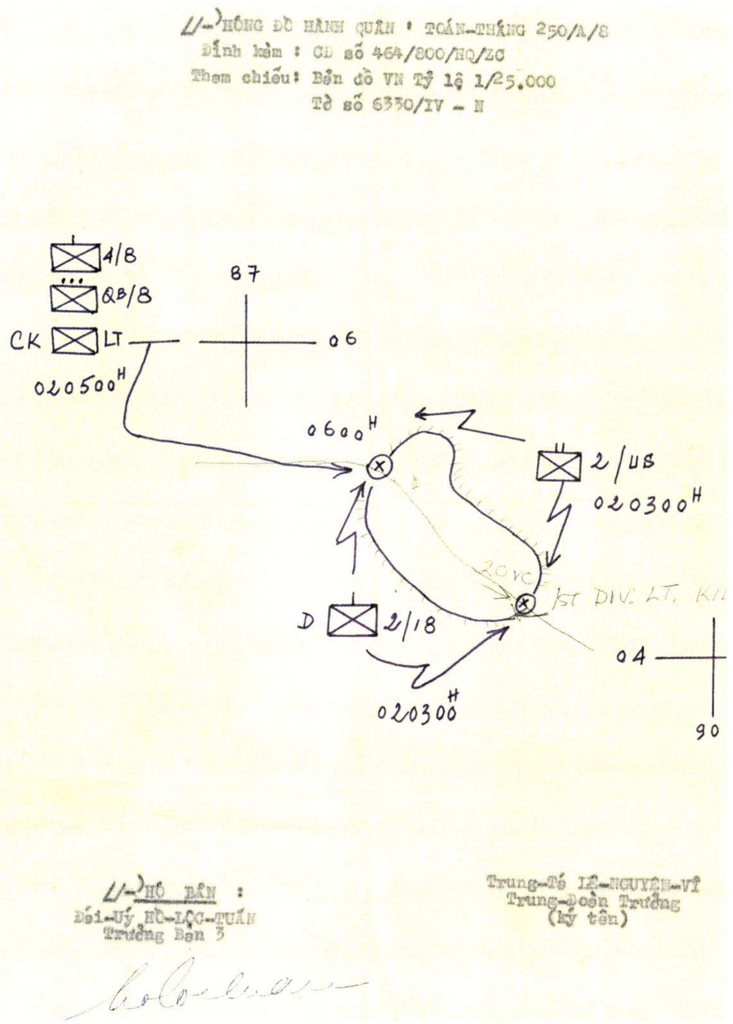

Another overlay, which consisted of a joint operation with the 5th ARVN Infantry Division's 18th Infantry Regiment and members of the U.S. 1st Infantry Division. (Carl Farren Collection)

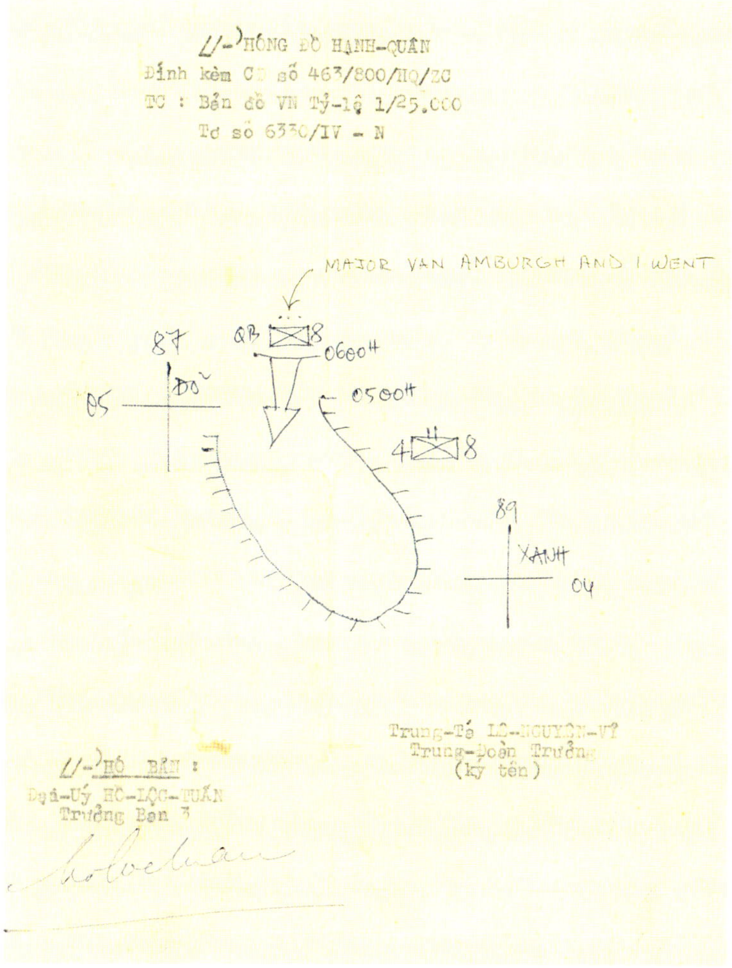

Copy of overlay that went over the map instead of writing on the map, in case the map fell into the wrong hands. Note the arrow and QB, which stands for RECON element of the unit in which Sp/5 Farren was attached. On this operation they were up against elements of the 9th VC Main Force Division. (Carl Farren Collection)

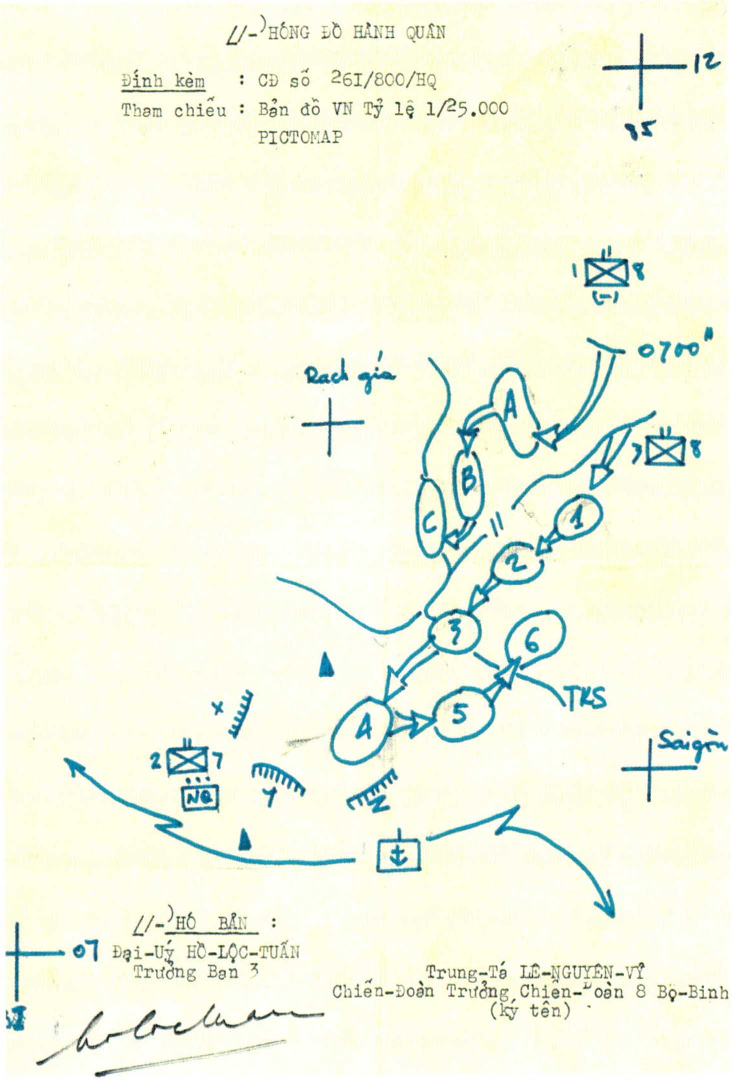

Overlay showing combined operations of the 5th ARVN Infantry Division's 18th Infantry Regiment. (Carl Farren Collection)

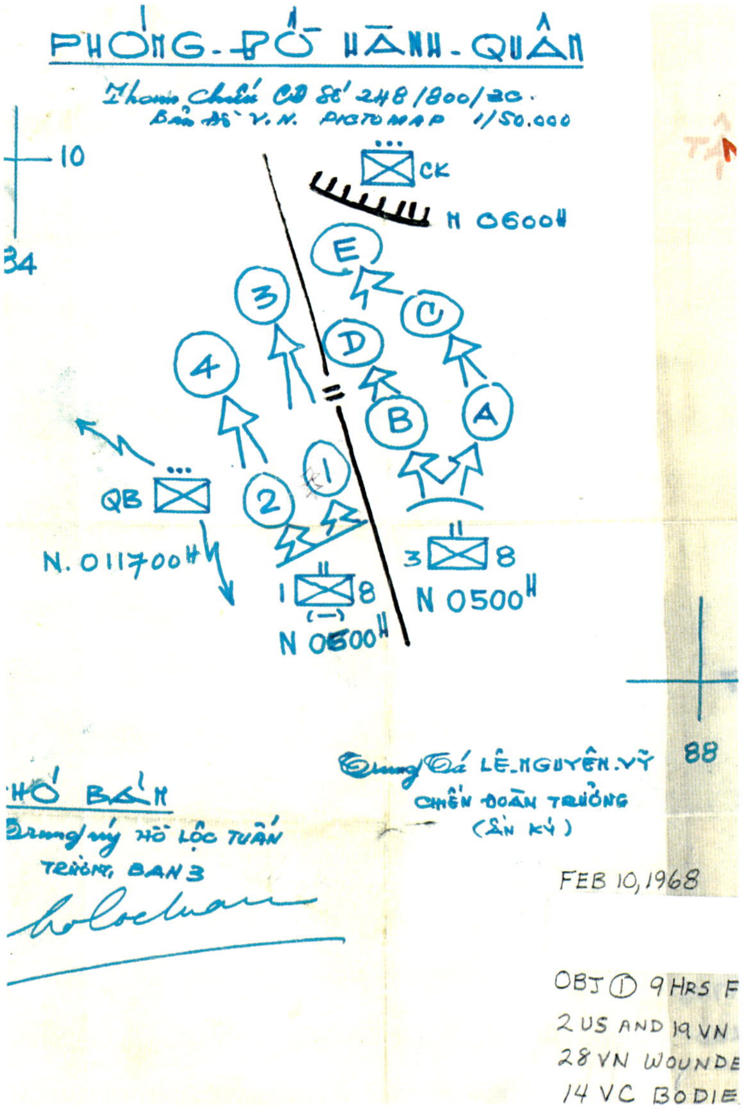

Another overlay which shows the operation conducted on February 10, 1968, and shows there were hours of fighting on the objective against the 9th Viet Cong Division. 14 VC bodies were found, although numerous blood trails estimated the count to be three times the number. This also shows in the lower right hand corner that two U.S. advisors were killed on this operation, along with 19 Vietnamese, and 28 Vietnamese wounded. (Carl Farren Collection)

Members of the ARVN 25th Inf. Div. during a ceremony honoring a fallen comrade.

**U.S. ARMY ADVISOR,
TEAM 99,
25TH ARVN INFANTRY DIVISION,
MR 3, TAY NINH, VIETNAM,
1967-68**

"In May 1965, 300 men were scattered throughout the ARVN 25th Division's tactical area, then a small detachment, where, according to Major Donald A. Siebert, living conditions were generally primitive. Mostly advisory compounds centered around a patchwork of French style Vietnamese buildings, prefabricated American structures, and various types of troop constructed shelters. Security was poor, and none of the compounds could have withstood a determined enemy attack. The operations center of the division detachment, supposedly the heart of the advisory team, was located in a gold Temple that had to be vacated at times for local religious services. Discipline was lax. When not in the field, team members' military duties were carried out in a leisurely fashion. They paid more attention to social functions than the war, and wore civilian clothing during off-duty hours. In manner and taste they reflected the slow unhurried tempo of the unit they advised."[67]

This advisor wears the third model poplin jungle fatigues with Vietnamese made E-7 (Sgt. First Class) on both sleeves, a crudely made MACV SSI on his left sleeve, and a Vietnamese made pocket

Close-up of crude Vietnamese made ARVN 25th Infantry Division pocket hanger.

Full figure view of U.S. Infantry advisor Team 99 to the 25th ARVN Infantry Division, Tay Ninh Province, 1967.

Full front view provides a good look at the advisor's M1A1 folding stock carbine.

Close-up of handbook for U.S. Forces in Vietnam.

hanger of the 25th ARVN Division. Around his neck is worn a red identification scarf made of colored cotton material. Different colors were given to each of the Infantry Companies in the battalion. Colors such as blue, red, pine green, and black were used to help the ARVN soldiers locate their company during combat and aid in link-ups. Worn on the right pocket flap is a 25th ARVN Division pocket hanger, which was usually worn by the advisors in this manner only. ARVN troops would wear this patch in the same fashion that the U.S. wore their SSIs on the left sleeve. Worn on the left sleeve of the advisor's jungle jacket is a very crude hand made subdued MACV patch. On each sleeve is worn his rank, E-7 Sgt. First Class, which were also locally produced. Above his left pocket is a Vietnamese made U.S. ARMY tape, along with a pair of U.S. made airborne wings. Web gear consists of an M1956 cotton pistol belt connected to the H-harness, with two universal pouches, two M1956 cotton canteen covers with the plastic canteen, and a single compass/first aid pouch, in which the advisor has secured the individual first aid dressing. On his H-harness and his pistol belt the advisor has secured two M18 red smoke grenades, most likely for marking targets or the LZ.

On his back he carries the items necessary for an extended operation in his ARVN Ranger pack. Most operations lasted less than 24 hours, so there was no need for a pack, but sometimes operations might last several days to a week, so many advisors would need a pack to carry their poncho and poncho liner, mosquito netting, hammock, water, food for the duration, ammo, spare batteries for the PRC/25-77, sun screen, and bug juice, as well as anything else he wanted to carry. Headgear is a locally produced cowboy hat in the traditional style made from U.S. parachute

Close up showing crude MACV pocket hanger with a U.S. made MACV SSI encased in plastic.

Right: *Upper torso shot provides a good look at the advisor's red identification scarf, worn by the different Infantry Companies for recognition on the battlefield. This also provides a good look at his gear.*

Close-up showing the advisor's ARVN Ranger pack.

Close-up showing the back side of the ARVN Ranger pack. Note the steel X-frame.

Close-up showing the side view of the ARVN Ranger pack. Note the numerous places for additional gear to be attached, either by the slide keepers or the M1910 hooks.

Close-up provides a good view of the advisor's subdued E-7 Sgt. First Class rank, and his hand sewn subdued MACV shoulder sleeve insignia. This also provides a good view of his ARVN rucksack.

Photo of South Vietnamese medic (Note the Unit medical bag) on patrol with members of the 5th ARVN Division, 18th Infantry Regiment. (Carl Farren collection)

Captain Orvis, on an operation in Tuy Hoa, wears the final model ripstop fatigues, along with red identification scarf, M1 helmet, and Mitchell pattern cover. The radioman next to him also wears the red company scarf, and carries the PRC-77 radio. (Carl Farren collection)

material. These were quite common and were comfortable and kept the hot tropical sun off the wearer's face, neck, and ears. Weapon is the handy M1A1 carbine with the folding stock. These were lightweight and compact, and were more manageable in the thick jungles and mangrove swamps that the soldiers had to operate in. They were also highly desirable to the other advisors. Third model jungle boots are worn here.

Close-up of infantry advisor's camouflage boonie hat. This one is done in the French style. (Owen Thorton collection)

2nd Battalion 18th Regiment RTO following behind Capt. Orvis in Tuy Hoa, 1969. (Carl Farren collection)

Additional close-up of Vietnamese made cowboy boonie hat. Note the 550 cord used for the chin strap, which is tied to the top of the hat. (Owen Thorton collection)

U.S. ARMY ADVISOR;
SOUTH VIETNAMSE 17TH RIVER BOAT COMPANY
ADVISORY TEAM 75;
MY THO
1967-1968

SSgt Fischer was assigned to Team 75 as Naval Advisor to the South Vietnamese Navy, due to his background in the U.S. Navy in Naval Intelligence. There were approximately four other Team 75 advisors at My Tho along with SSgt Fischer: the commander, LT. Col. Showalter, a NCO who was the medic, also known as Bac Si, Tom Woods; an Intelligence officer: and an ARVN Armor Advisor. Province Headquarters was at Dinh Tung, and was commanded by a LTCR. Team 75's mission was to train five districts in which they supervised both Navy and Army, and to interdict VC and supplies coming down the river from Cambodia. Other times SSgt. Fischer would accompany a pair of boats three miles up river and set an ambush on one of the smaller canals. When they returned from ambush duty, the Advisor would hold inspection, dismiss sailors for chow or rest, and grab some sleep himself before going out on ambush again the next night. Team 75's

My Tho air strip, Aussi caribou landing on short runway, 1967 (courtesy JC Fischer collection)

Staff Sgt. J.C. Fisher wearing the third model poplin jungle fatigues, pin-on E-6 rank, a subdued MACV SSI, and a hexagon shaped Vietnamese Navy pocket hanger. The Vietnamese Naval NCO standing next to Staff Sgt. Fisher wears the traditional Vietnamese Naval uniform. Note the rank on his collar. My Tho, compound of the 17th River Boat Company Intel., 1967. (Courtesy JC Fischer collection)

J.C. Fischer center wearing third model jungle fatigues with the Vietnamese made ultra Marine blue Naval beret. Note the large E-6 stripes on his sleeves and the shield shaped Vietnamese Naval pocket hanger. The Vietnamese from left to right are the 17th NCOs and unknown Officer, Capt. Tuong, and their 1st Sgt. All the Vietnamese wear U.S. style OG 107 cotton fatigues with Naval fatigues caps. Note the two on the right wear the naval beret badge on the front of their hats. The 1st Sgt. (far right with the sunglasses) wears the Vietnamese Navy patch on his left sleeve. Compound of the 17th River boat company Intel., My Tho, 1968. (Courtesy JC Fischer collection)

Team house for advisory team 75. This was once a French plantation owner's home, My Tho, Dihn Toung province, IV Corps., 1967-68. (Courtesy JC Fischer collection)

Close-up provides a good look at the ultra marine blue Vietnamese Naval beret.

Upper torso shot of a team 75 advisor to the 17th River Boat Company Intel Section, My Tho, 1967-68.

Upper torso provides a good look at the advisor's yellow Vietnamese Naval pocket hanger.

Upper torso shot provides a good look at the advisor's blue Naval beret with Vietnamese Naval device. This also provides a good look at the Vietnamese made E-6 stripes and hand sewn CIB. (Courtesy of JC Fischer)

Staff Sgt. Fisher (left) with Capt. Toung (right), the Commanding Officer for the 17th River boat Company Intel, My Tho, 1968.

Additional close-up provides a good look at the naval device and the locally made tailor tag in the beret.

Advisor compound was once a lush French Villa at one time, and now served as H.Q.

Uniform worn here is the 3rd model poplin "Jungle fatigues" worn with a Vietnamese made arc that reads "VN Navy Advisor." This was handsewn on OD material and purchases in the town of My Tho. It was designed to be worn over the MACV SSI, and is another example of unauthorized patches being worn by American advisors. Worn above his left and right shirt pockets (obscured by his Flak vest) are a hand sewn U.S. ARMY, CIB, and nametape. Worn on the left pocket is a yellow shield with a black anchor, symbolizing the Vietnamese Navy. During missions, headgear consisted of a black stocking cap, sometimes referred to as a "watch cap," a tigerstripe boonie hat, or OD boonie hat. Helmets were routinely taken along but seldom worn. Flak jackets were unseen until the "TET" offensive, when everybody got one. Side arms consisted of a .38 cal pistol that was usually carried in a shoulder holster or in one of his pockets. SSgt. Fischer also carried on occasion a Russian/ChiCom Tokeroff pistol that was a captured item. Weapon is the early model Thompson submachine gun with the short magazines. Additional magazines could be carried in his lower shirt pockets, the pockets of his M69 flak vest, or in his large Musset bag, sometimes referred to as a "ditty bag." The large OD green WWII era bag was compartmentalized and could accommodate several of the short or long Tommy gun magazines, several M26 hand-grenades, or the M18 smoke grenades. Two additional red M18 smoke grenades are carried connected to the outside of his flak vest and in the pocket. Note also the "Skill craft" pen secured in the grenade loop. Jungle boots are the third model with the DMS.

U.S. ARMY ADVISOR;
SOUTH VIETNAMSE 17TH RIVER BOAT CO.
ADVISORY TEAM 75
MY THO
1967-1968

This recreation of SSgt. J.C. Fischer shows what the Naval advisor wore around the compound and at the schoolhouse at My Tho. The shirt is the 3rd model poplin jungle fatigues with a handsewn U.S. ARMY tape, a very crude handsewn CIB, a U.S. made twill MACV SSI, and Vietnamese made E-6 (Staff Sergeant) chevrons, which are lined with black felt.

Two American advisors, along with several South Vietnamese Naval officers during inspection of the school at My Tho, 1967-68. (Courtesy JC Fischer collection)

Connected to the left pocket flap is a very unique pocket hanger, which features a black anchor with a rope against a yel-

Full figure shot of Team 75 advisor to the 17th River Boat Company prepared for night operation on My Tho River, 1967-1968.

Full rear shot provides a good look at the advisor's carry all bag and M69 flak vest.

Full figure view of advisor with emphasis on his right side. This also provides a good look at his carry all bag and early Tommy gun.

J.C. Fischer being awarded the Cross of Gallantry with Palm and Vietnam Navy Service Honor medal by Capt. Cat. Note the fresh coat of paint, the new grass, and white painted bricks and gravel walkway—J.C. Fischer started all improvements. 17th River Boat Company's school house and classroom, My Tho, 1968. (Courtesy JC Fisher collection)

J.C. Fischer shaking hands with Captain Cat after being awarded two medals. Note the Vietnamese sailors all wear U.S. style fatigue caps and U.S. style OG 107 sateen uniforms. The 17th River Boat Company Intel., My Tho, 1968. (Courtesy JC Fischer collection)

Close up of Team 75 advisor. Note the watch cap, M-18 smoke grenade, and Tommy gun.

Upper torso shot provides a good look at the advisor's early model Tommy gun. Note the fullness of his lower two pockets, which are full of M26 hand-grenades.

low background. Another variation worn by Team 75's Navy advisor was slightly different. The shield was a hexagon, and was also in the pocket hanger device. Other variations featured a small black number at the base of the patch. The pocket hanger is a printed variation encased in plastic.

Headgear is the ultramarine blue "Hai Quan" Naval beret with the naval device. The beret device is handsewn and features a white anchor surrounded by two yellow wreathes. SSgt. Fischer was the only one at My Tho who wore the blue beret. The VN Naval personnel, including their commanders, all wore the Marine/Navy style fatigue hat, or "stop sign" hat, with a black VN Naval device sewn to the front. Some officers had the "scrambled eggs" on the bill. All photos and correspondence with SSgt. Fischer confirm that he was the only one wearing this unique blue VN Naval beret, which was awarded to him at a ceremony. The blue beret is one piece, and is in the "French style," being very large on top. This beret was rarely worn even around the compound, due to the heat of the tropical sun. Most of the time, a boonie hat was worn during the day, and at times on operations at night. The beret would never be worn onboard the boats, and especially not on operations.

Uppper torso shot provides a good look at the locally produced VN NAVAL ADVISOR arc, which was usually worn over the MACV patch.

Full figure shot of American advisor to the 5th ARVN Division Recondo Company, Operations against the 33d NVA BN. In Phu Hoa Dong, February 21, 1968.

U.S. INFANTRY ADVISOR
5TH ARVN INFANTRY DIV. "RECONDOS"
TEAM 70,
Co Lam Son,
Operations in the Iron Triangle
1967-1968

Uniform consists of a set of the MDAP contracted tigerstripe in the traditional pattern. Note this one has been modified into a short sleeve garment. His tigerstripe boonie hat is a near match, probably purchased at the same tailor shop. Around his neck is worn the RECONDOS identification colored scarf—black. These were nothing more than colored pieces of cotton cloth made into scarves for the different companies.

Web gear consists of two "universal pouches," sometimes mistakenly referred to as "M-14 pouches," and two canteens with covers. The two "Universal, small arms pouches were designed to carry a wide variety of magazines. It could hold two B.A.R. magazines, one folded bandoleer of M-1, six M-1 clips, or four 30 round carbine magazines. The pouch closed with a quick release fastener by putting a web tab through a billet sewn on the front of the pouch. The case also featured a back-supporting strap that could be extended and snapped to the front of the suspenders, which kept the pouch from hanging forward when loaded

Full figure shot of RECONDO advisor with emphasis on his left side provides a good look at his M-17 mask and his lightweight rucksack.

Full figure shot of 5th ARVN Div. "RECONDO" advisor, Team 70, 1968.

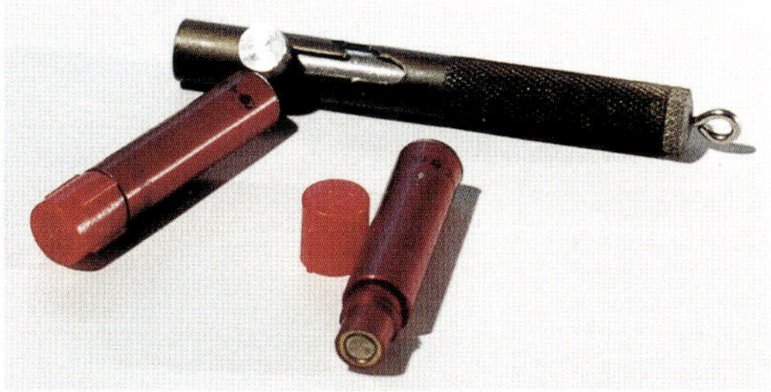

Close up of mini-flare launcher carried by RECONDO advisor while on operations.

with ammo. Grenades could be secured on either side by means of a strap and slots."[68] Each American advisor was usually issued two of these pouches. The canteen covers, like the rest of the components of the M1956 gear, are made of cotton duck in OD number 7. It is lined with grey wool felt, which when wet is designed to keep the water in the 1-quart canteen cool. Worn on the left side of his pistol belt is the M17 protective mask in its cover. The cover is also made of cotton duck material in OD shade 7. The carrier attaches to the wearer's waist with a strap, and an additional strap secures the bag to the leg to keep it from bouncing around while moving. A single M18 smoke grenade is attached to the left side of his H-harness, along with a compass/first aid pouch. Connected to the right side is a pair of the Army issue sunglasses, which were very crude and consisted of plastic tinted lenses on a flexible wire frame.

Close-up showing the lightweight rucksack.

Close-up showing the backside of the lightweight rucksack.

Close up of 9th ARVN Infantry Division "TS' RECONNAISSANCE.

Close-up of two South Vietnamese patches. The one on the left is the 3rd ARVN Infantry Regiment, Reconnaissance Company (spider man). This one is the printed variation, and it features the legendary super hero from Marvel comics. The one on the right is a silk woven variation from the 2nd Battalion, 11th Infantry Regiment. The words "Kim-Long" at the bottom are Vietnamese for Yellow Dragon.

RECONDO advisor kneeling provides a good look at the lightweight rucksack.

RECONDO advisor kneeling provides a good look at his left side and his M17 protective mask.

Close-up showing U.S. Army soldier's notebook, available for purchase at the PX. Inside would be areas for commander's name and other information.

Unknown MACV advisor sits on his bunk holding his CAR-15 rifle. Note the shark's tooth around his neck. Circa 1968.

"The lightweight rucksack is seen here with its nylon bag secured to the bottom of its aluminum frame. The Lightweight ruck could accommodate twice as much as the M1961 buttpack, was water repellent, and featured double eyelet web tabs for the attachment of other items, such as a 1qt canteen with cover. First produced for Special Forces as the T62-1 in 1963, it was originally intended for arctic use. The ruck was standardized on November 1965."[69] Weapon is the M16A1 rifle with a single twenty round magazine. Lt. Parrish was able to obtain enough of these rifles to outfit the entire Recondos.

Close-up showing the RECONDO's tigerstripe pants. Note the white label that reads A M, for Asian Medium. It was always a challenge to obtain tigerstripe uniforms in "human sizes."

Close-up showing the advisor's short sleeve tiger stripe shirt.

Two Vietnamese soldiers from the 5th ARVN Infantry Division. Both are armed with the new M-16 rifles, and they both wear uniforms similar to U.S. style OG 107. Circa 1968

U.S. INFANTRY ADVISOR, 1ST ARVN INFANTRY DIVISION, "STRIKE FORCE" COMPANY, MR-1 Thua Thien, Vietnam 1968-1969

The camouflage utilized by the ARVN Rangers, LLDB, Reconnaissance Companies and their U.S. advisors was a direct descendent of the U.S. ERDL pattern. Starting in 1962, the Advanced Research Projects, Agency Research Development field unit (ARPA) began production and testing of a new reversible combat uniform for the Vietnamese. These test uniforms provided an ERDL leaf camouflage pattern on one side and black on the reverse side to resemble a pair of V.C. pajamas. By late 1964, the Biet Dong Quan had adopted the pattern, and it became known as the ARVN Ranger pattern, although other branches, such as the Airborne and the LLDB, wore and utilized this pattern, also. Although this pattern was being recognized as the Ranger Pattern, some ARVN Recondo units continued to outfit their personnel in the Duckhunter pattern and/or the Tigerstripe uniforms.

The Ranger pattern, worn here, was produced in a variety of colors and shades on different weights of material, but was still true to the original design. There has been some debate on which came first and where the variations were produced. CSM Mike Martin, author of Black Tigers, remembers the lightweight uniforms first appearing in 1965 while he was serving with the 44th

Full figure view of U.S. advisor to the 1st ARVN Infantry Division, Reconnaissance, Strike Force Company, MR 1, Thua Thien Province, 1969.

Full figure shot with emphasis on the advisor's left side. This provides a good view of the advisor's MACV patch and his ARVN Ranger pattern fatigues.

Full figure shot of advisor with emphasis on his right side. This provides a good look at the tan Vietnamese Infantry beret.

Full figure view of RECONDO advisor provides a good look at his tan Vietnamese Infantry beret.

Full figure view of the advisor "taking 5" while waiting to move out on the operation.

ARVN Rangers in the Delta. By late 1968 they resumed printing the distinct pattern on the heavier cotton material. It is not clear if the production of the lightweight poplin uniforms ceased or continued in smaller quantities. By 1969, it seems the heavier weight Ranger pattern was predominant. Worn on the advisor's uniform is a subdued twill MACV SSI, a subdued U.S. Army and matching printed name tape, green leadership tabs on the epaulets, a subdued 1st Lt bar on his right collar, and crossed rifles on his left collar, along with his Vietnamese rank pinned to the front of his shirt and a printed variation of the 1st Division Strike Company. "The patch features a red disc, with a narrow black border, a Black Panther with green eyes, and the words 'Sat Cong' (Kill Communist)." Around his neck the advisor wears the strike company's distinctive scarf color of black, though sometimes a piece of camouflage para silk was worn, too. These were worn for identification purposes, and each company had its own distinctive colors.

Headgear is the tan Vietnamese Infantry beret, worn with the gold officer's beret badge, and is the correct size for the beret—not the cap badge being utilized as a beret device like we've seen before! Next to the beret device is the officer's rank, Trung Uy (Lieutenant), which was usually worn on the berets.

Web gear consists of an M1956 pistol belt, cotton H-harness, two universal pouches, one with the compass/first aid pouch connected to it, and a single M1956 canteen cover with a newer 1-quart plastic canteen, which were first introduced to troops in 1963."[70] Worn on his left hip is the black leather holster for his .45 pistol. Worn on the rear of the pistol belt is the M1956 Combat Field pack, or "Butt Pack," which was designed for mobile armored infantry operations and supported by his suspenders. Inside would be carried extra socks, grenades, rations, and magazines. The advisor utilizes the ARVN ruck, in which he would carry extra ammo., his rations, water, a hammock, poncho, and poncho liner. Attached to the outside of the pack is a single universal pouch and M1956 canteen cover with an additional plastic 1 qt. Canteen.

This advisor wears the second model DMS boots, which like the first model lacked the ankle reinforcement strip, which would be incorporated in late 1965. The second model Tropical Combat Boot, DMS, did away with the leather strip running up the Achilles' tendon and the leather strip along the top of the boot. A new type of slip-in spike resistant insole was available, which was strips of steel encased in plastic, but many soldiers and advisors alike decided not to utilize them due to the fact that when worn on long marches the insole got hot and caused unnecessary blisters. This style, the second model Tropical Combat Boot, or "Jungle boot," were the first ones issued in any great numbers. There were several earlier variations without the DMS, but these were few and far between, and for all intents and purposes this is the true 2nd model that was adopted by both the Army and Marine Corps in 1965. Weapon is an M-1 carbine without a strap, and the smaller magazine.

Upper torso shot provides a good look at the advisor's black recon identification scarf and his M1 carbine.

Upper torso shot provides a good look at the advisor's SAT CONG patch, ARVN rucksack, and black holster for his .45 pistol.

Close-up of printed 1st ARVN infantry division, Strike Force Company patch.

Left: *Upper torso shot provides a good look at the advisor's ARVN Ranger pattern camouflage with the Sat Cong pocket patch of the first infantry strike force company. This also provides a good look at his U.S. Army tape, name tape (Wolfe), U.S. airborne wings, and green leadership tabs on his epaulets. This also provides a good look at his tan Vietnamese Infantry beret. Note the rank of Lieutenant pinned next to the beret badge.*

Full figure view of advisor without his gear and rucksack. Note the two patterns are the same, but the trousers have a more green tone, while the shirt has a slightly blue tone to it. It was not uncommon to see Americans or Vietnamese with mismatched colors like this.

Close-up showing the tan Vietnamese Infantry beret with the correct size gold officer's beret badge.

Close up of gold ARVN officer's Infantry beret badge (left) with the ARVN rank (right) of Trung Uy, 1st Lieutenant.

MAT TEAM ADVISOR, TEAM, 96 TUY HOA, Vietnam 1970-71

"The province Territorial Forces advisor, normally an Army Captain, and the small district advisory teams (still about 4 man each) assisted the territorial units as best they could. The small advisory cells also operated with Popular Forces (PF) training centers, the territorial logistics companies, and the smaller specialized units. But the Territorial Force advisory effort was too small to have much of an impact, and advisors spent most of their time working behind the scenes to make the weak territorial command, administration, and supply system work. The South Vietnamese regular army continued to support the territorial with some training and supplies, but showed little interest in their deployment and operations."[71]

"Following the creation of CORDS in April 1967, Westmoreland set about to strengthen the area security advisor organization. During the subsequent months he established Mobile Advisory Teams (MATs) to work directly with the small ter-

Sign outside of the Tuy Hoa district headquarters. This is where Team 96 lived and worked. (Courtesy Don Driftmier collection)

Numerous MACV advisors at the advisor's school in Long Binh. The barbed wire in front of them is a Vietnamese tiger cage. Note the pocket hangers worn by the two men in the center. (Courtesy of Don Driftmier)

2nd Lieutenant Don Driftmier on the left with unidentified Lieutenant, standing in front of the U.S. Army Vietnam, USRAV, Advisory School Library, Long Binh, Vietnam, 1970. Both men wear the U.S. Army Advisory School patch, worn by students and the cadre alike. Students would return the patch after completing the school. The patch was the same shape in size as the MACV SSI and was divided into thirds. The top left featured the coat of arms of the regional force/ popular force, the top right featured the popular force emblem, and the bottom featured the MACV emblem. Both Lieutenants wear rip stop jungle fatigues, jungle boots, and baseball caps. (Don Driftmire collection)

Headquarters building of the U.S. Army Vietnam Advisory School in Long Binh, 1970. (Don Driftmier collection)

Vietnamese officers stand in front of the flag pole outside of advisor's compound, Tuy Hoa, 1970. (Don Driftmier collection)

Lieutenant Don Driftmier stands with Vietnamese soldiers outside of advisor's compound, Tuy Hoa, 1970-71. (Don Driftmier collection)

Lieutenant Don Driftmier (standing) watches as a member of the revolutionary development cadre—in the black pajamas—instructs regional force members in all weapons, Tuy Hoa, 1971. (Don Driftmier collection)

ritorial units in the field, Mobile Advisory Logistics Teams (MALTS) to assist the territorial logistics companies in each province, and deployed several hundred MAT and MALTs by the end of the year. Under the overall direction of a province senior advisors, each team consisted of two junior officers and three to four enlisted men. At the same time the MACV commander integrated similar teams, formed by American tactical units, into the field advisory organization, and also enlarged the size of the province advisory teams by adding training, engineer, and personnel advisors. In 1967 the number of CORDS advisors for the provinces, districts, and Territorial Forces thus increased from about 2000 to over 5,000, while the strength of the tactical, or combat advisors, remained stable at about 3000."[72]

Lt. Don Driftmier with Vietnamese soldiers outside of village in Tuy Hoa, 1970. (Don Driftmier collection)

Lieutenant Don Driftmier's cot with mosquito netting, along with his collage of beer, a car, and girls. Advisor's compound, Tuy Hoa, 1971. (Don Driftmier collection)

MAT Team advisor lays on bunk, advisor's compound, Tuy Hoa. (Don Driftmier collection)

Don with pink flashlight, rip-stop jungle fatigues, and boonie hat on operation in Tuy Hoa, 1970. (Don Driftmier collection)

"In 1967 it was suggested that MAT Teams increase strength of the advisory teams assigned to Vietnamese battalions performing territorial security to provide an advisor for each company. As an alternative, the MACV staff proposed assigning two extra men, an experienced NCO (E-7) to each 4 man battalion advisor he teamed serving with a security tasked battalion. However, considering the shortage of experienced personnel, General Westmoreland decided against either proposal, and the matter was dropped.

By the spring of 1967, the quality, experience, and morale of the advisors had also become a serious issue. In May, Westmoreland again asked Gen. Johnson, the Army Chief of Staff, to assign the Army's best officers to American tactical units in South Vietnam (that is, the Army component command), and not to MACV and the advisory network. The new MACV Deputy Commander, Gen. Abrams, felt that Gen. Westmoreland's priorities were inevitable, yet he was worried and communicated his concerns to the Army staff. Responding to Abrams, Johnson confided that he was doing everything possible in Washington to enhance the status of the advisors, but had to give priority to the requirements for troop commanders, noting that the U.S. Army was a 'fighting force,' and the ability to command troops effectively was its primary measure of success. General Abrams responded that the advisors not only saw themselves as 'second class citizens,' but also were treated as such, and felt that it was necessary to make a concerted effort to increase their status in their own eyes."[73]

"Noting the increased attention given to area security, Gen. Johnson felt that he could best help by improving the quality and incentives for prospective province senior advisors. Under Johnson's prompting, the Army Staff upgraded standards for these posts and sought to identify qualified candidates several months prior to their assignment to Vietnam. Nominees were to be superior combat arms officers with a rank of colonel, or promotable

Full figure view of Mobile Advisory Team (MAT) advisor, Team 96, Tuy Hoa, 1970-71.

Front view provides a good look at the gear utilized by MAT team advisor in 1970.

Another view of the team 96 advisor provides a look at his ARVN rucksack and the newer short M-16 ammo pouches.

Lieutenant Colonel who had commanded units in Vietnam; they were also to be graduates of the U.S. Army War College, with the ability to speak or an aptitude for learning Vietnamese. With General Westmoreland's approval, the entire package included advanced courses at the U.S. Foreign Service Institute, extensive Vietnamese language training, 18 month tours, and generous leave privileges, extra pay, and special consideration for next assignment. Westmoreland and Johnson also agreed to award command credit, an important factor in promotion consideration, to all province and district senior advisors. General Abrams, for his part, supported the establishment of a special advisory school in Vietnam for the future mobile advisory team members. Approved October 1967 and staffed by handpicked advisors from each zone, the school began concentrated two week training sessions in early 1968. Abrams also insisted that only highly experienced personnel be assigned to the territorial advisory program, and when the Army Staff was unable to fulfill his requirements, he had the MACV and U.S. Army Vietnam staffs transfer personnel who had the requisite qualifications directly from American combat units in Vietnam."[74]

Don Driftmier prepares for operation with his jeep at the advisor motor pool. (Don Driftmier collection)

Special Forces trained medic stands outside of Vietnamese hospital in Tuy Hoa. He wears a set of rip stop jungle fatigues, jungle boots, and tiger stripe boonie hat. In his left hand he carries an M-79 grenade launcher. (Don Driftmier collection)

"Hardly any advisors spoke Vietnamese, and their communications with counterparts depended heavily on native interpreters. Departures of advisors without any replacements continued to be commonplace, damaging the continuity of the effort at the lower echelons. Under pressure to project a positive image of the South Vietnamese Army, many advisors continued to avoid passing on adverse comments to their superiors. The problems were not new, having existed throughout the advisory network for years. But now they had become severe in the small battalion and regimental advisory teams, where the advisors were normally inexperienced young officers who vetoed advisory duty as detrimental to their careers. On the battlefield, field advisors continued to play their combat support role. Whether the South Vietnamese were performing 'search and destroy,' clearing, or securing missions, the requirements and tasks at the dirt level never really changed throughout the war. In the 'boondocks' on operations, the battalion advisor needed a good pair of legs and strong back; water, and perhaps a bit of food (rice); a variety of skin lotions (suntan, mosquito repellent, foot powder, and medicinal creams); a pack of cigarettes; hard candy or chewing gum; and a good map, an accurate compass, and reliable radio with extra batteries. His rifle, a few clips of ammunition, smoke grenades, and assorted odds and ends (pen and paper, some kind of small knife, rope, extra stocks, a towel, and perhaps a good paperback book)

Full view with emphasis on the advisor's right side. This provides a good look at his M1911 A1 45 pistol in the black leather holster and U.S. Army Vietnam shoulder sleeve insignia showing a combat tour.

Rear view showing the MAT Team leader's ARVN rucksack with PRC-77 radio; an essential piece of gear for extended operations.

completed his basic needs. The less he carried, the better. Most operations involved long, difficult treks through overgrown marshes and jungles in stifling heat and humidity. Boobytraps and mines were his worst enemy, with fatigue, insects, and disease ever-present but minor irritations."[75]

This MAT team advisor wears the new rip-stop jungle fatigues, which were basically cotton-poplin and incorporated a nylon weave which greatly increased the strength of the fabric. On the left sleeve he wears the late model subdued MACV patch, which was U.S. made and featured a marrowed edge. On the collar points (obscured by his flak vest) are pin-on crossed rifles and 1st Lt. Bar. Worn above his nylon U.S. Army tape is a subdued pair of airborne wings and a CIB. On the right is a Vietnamese made name tape.

"The AN/PRC-25 was a short-range man pack portable, frequency modulated (FM) receiver/transmitter, weighing just over 24 pounds with the dry cell battery. Each battery had an average life of about 20 hours. Maximum range for the radio was from 5 km with the short tape antenna and up to 8 km with the long seven section "fish pole" antenna. The radio was part valve/part transistor and had 920 channels, in which two were pre-set."[76] Note the white plastic spoon has been added to allow the advisor to secure the hand mike to his suspenders. The radio is seen carried in the ARVN Ranger pack. The ARVN ruck featured its own

Full view showing the MAT advisor utilizing the PRC 77 radio. Note the rough condition of his M18 smoke grenade.

Don Driftmier and two other Americans from Team 96 prepare for operations. Note the medic on the left wears the M79 grenade vest, along with the M-79 grenade launcher. (Don Driftmier collection)

Three members of Team 96. Note the man in the center wears the experimental load bearing vest for the 20 round magazines for his M-16A1 rifle. The man on the right wears a M1951 patrol cap and carries a twin cell pouch usually seen with the Air Force or M.P.s. (Don Driftmier collection)

MAT team advisor utilizing the radio. This also provides a good look at the advisor's right side and his pink flash light.

1st Lieutenant Driftmier wearing a set of rip stop fatigues with a nylon U.S. Army strip, hand sewn name tape, machine made MACV SSI, and sewn on collar rank and BOS Crossed Rifles. He sits inside a hootch with one of the team's many puppies. Tuy Hoa, 1970. (Don Driftmier collection)

Numerous MACV advisors from MAT Team 96 enjoy a traditional meal with tea at the Providence Chief's house in 1970-71. (Don Driftmier collection)

MAT Team advisor sits down to enjoy a movie in the day room of the Tuy Hoa MAT compound. (Don Driftmier collection)

Lieutenant Driftmier's refrigerator, which is Holly® equipped, along with personal items, such as books and a picture of his wife, inside his room. Note the bottle of brandy that sits next to the refrigerator. (Don Driftmier collection)

MAT team advisor kneeling down. This provides a good look at the advisor's pink flashlight and ARVN rucksack.

Team 73 MAT Team advisor patch.

integrated spring steel X-frame, with a rivet placed uncomfortably in the center. Many advisors and ARVN soldiers took the X-frame completely out or bent it inwards to avoid the rivet, which was strategically placed at the center of the wearer's back. The ruck also featured two smaller pockets on the outside and numerous places where the M1910 double hook system or the M1956 gear could be added. Inside his ruck he carries an additional battery, the day's rations, and personal gear.

Web gear consists of a M1956 pistol belt with no suspender. Attached are two of the newer canvas Small Arms Ammunition Cases, M16A1 20 round. First introduced in 1968, these were designed especially for the 20 round magazine. Attached to one is a single "Baseball" Hand-grenade. Worn on either side of the ammo pouches are a single M1956

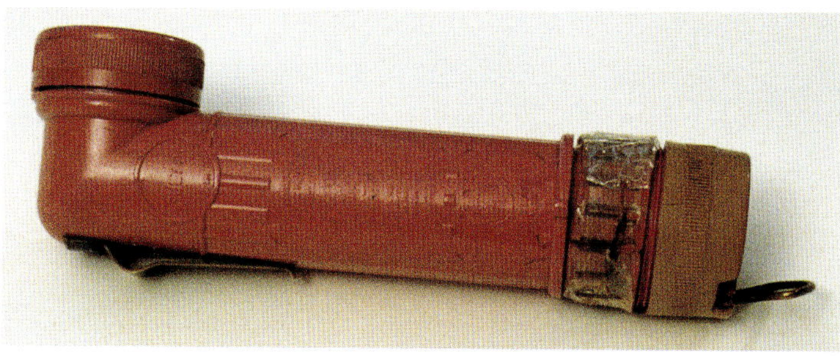

Close-up showing the side of the pink safety flashlight.

Additional close-ups showing pink safety flashlight.

Right: *Vietnamese interpreter wearing the ARVN Ranger pattern camouflage with the matching Marine Corps style fatigue cap. Note he wears his ribbon bar pinned above his left pocket. He sits inside the MAT team's day room, invited in by the Americans to watch a John Wayne film. Note the fire extinguisher mounted on the wall, and the brass vase with plastic flowers on the dining room table. (Don Driftmier collection)*

canteen cover with a plastic 1qt canteen, and on the right is the black leather holster for his .45 pistol.

"The Body Armor, Fragmentation Protective Vest, M69 is worn here, and was an upgraded version of the earlier M1952, and derived its protection capabilities from layers of ballistic nylon, twelve in front and upper back, two in the lower back, and a further two down the length of the spine. The M69 omitted the shoulder loops of the M1952, but incorporated a new three-quarter collar. The first production vest, seen here issued prior to the 1968 TET offensive, had a zippered closure. Later types had Velcro fastening, and both versions could be adjusted to fit by laced closures on the side."[77]

Revolutionary Development Cadre member wearing black pajama style uniform with the rural development cadre patch on his left pocket. This patch featured a yellow map of Vietnam surrounded by two heads of rice, which were also yellow against a green background with a white border. The words TO-QUOC (country) and NHAN-DAN (people) are also white. These were both silk woven and printed. The soldiers would show up on motor bikes, teach a class, then disappear. Tuy Hoa, 1970-71. (Don Driftmier collection)

Vietnamese Lieutenant sits inside of Don Driftmier's room prior to Lieutenant Driftmier's departure for the U.S. He wears Vietnamese made Army fatigues, styled after the U.S. OG 107 cotton sateens. His Web gear consists of an M1967 H-harness in nylon with a matching M1967 pistol belt. Note the Korean War era first aid pouch and U.S. black holster. Behind him in the upper right corner is the ARVN Lt.'s farewell gift to Lt. Driftmier, which consists of several large shells and Coral, 1971. (Don Driftmier collection)

Close-up of two different Regional Force/Popular Force patches. The one on the left is a pocket hanger variation which features a paper device suspended in plastic, while the one on the right is a printed variation. Both feature the suit of armor and sword.

Connected to the right side of the vest is the MX-212A4-U safety flashlight. This was made by Brightstar Bulb, and was a Type S (Safety Class 1-Group D) PERMISSIBLE ELECTRIC FLASHLIGHT. These had a non-sparking bulb, and were used aboard aircraft, ships, tunnels, or anywhere that may contain flammable gases that the regular flashlight could ignite. On the left of his M69 is a single M18 smoke grenade.

Headgear is the standard M-1 steel pot without the camouflage cover. Weapon is the M16A1 5.56 rifle with a single twenty round magazine.

Additional close-up showing the side of the camouflage baseball cap.

Typical of the Vietnamese made camouflage baseball cap, this one is done in poplin ERDL camouflage pattern. These were worn by members of the RF/PF.

Close-up showing the inside of the baseball cap. Note the blue liner and the plastic sweatband in the front.

Lieutenant Driftmier's cot, along with a camouflage poncho liner, his helmet, boonie hat, box fan, locker, and weapon. Note the primitive living conditions, 1970-71. (Courtesy of Don Driftmier)

Two advisors take time to catch up on some reading before the movie starts inside the MAT team's day room. Note the large selection of magazines at the magazine rack, everything from Ebony to Popular Science, Time, and numerous "muscle car" magazines. (Courtesy of Don Driftmier)

3

U.S. Airborne Advisors Team 162 1964-1971

"The Vietnamese Airborne began in 1946 when the French Expedition Area Corps in Indochina organized the very first Vietnamese airborne unit. It was company sized, based on a similar organization of the British SAS (Special Air Service). The United States Army filled the vacuum created by the FEC's departure from Indo-China following the Geneva Convention of 1954. A Military Assistance Advisory Group (MAAG) had already been established in September of 1950 to coordinate the flow of U.S. aid, but now it became the dominant military presence in the South. Meanwhile, the new president, Ngo Dinh Diem, quickly consolidated his power in the new republic by using the airborne battalions to destroy the 'Cholon Mafia,' known as the Binh Xuyen. These gangsters had operated with impunity in the Chinese quarter of Saigon throughout the colonial period; Diem used his paratroopers to rid the city of this threat. His handling of this crisis reinforced U.S. support of the new government and caused the MAAG chief to make plans to establish an advisory detachment with the Vietnamese Airborne Group. The creation of the Vietnamese Airborne Advisory Detachment, later known as MACV Team 162 and called the "Red Hats" by most, began an 18 year love affair that finally ended on a hot February afternoon in 1973 with the deactivation ceremony, prompted by the January cease-fire agreements.

Members of Team 162, Vietnamese Airborne Brigade, Advisory Detachment 1964. Note the wide variety of uniforms worn by these early advisors. The advisor at the bottom left of the photo wears a modified U.S. first model exposed buttons shirt, made into a two pocket shirt and tucked into his first model exposed button trousers. He has also had the pockets straightened out. The Lieutenant, at the far right of the photo, holds a clipboard and wears a set of the Vietnamese Airborne pattern camouflage, known as the "Bullet Proofs" by the advisors. Note the nametape on his left side and the maroon Airborne beret. Also note the advisor in the far left, back row, wears a pop-up Ridgeway style hat with wings on top and what appears to be either Major or Lieutenant Colonel rank on the bottom. All three advisors are wearing the "Bullet Proofs" pattern camouflage, and all wear their name strips on the left side; this is usually only seen with Airborne advisors of team 162. (Courtesy Mike Martin)

The initial advisory efforts with the Vietnamese Airborne Group were technically oriented rather than tactical. They emphasized planning, training, equipment, and logistical issues. A reflection of growing U.S. Army staff influence emerged in 1959 when a Joint General Staff (JGS) redesignated the Group as the Vietnamese Airborne Brigade. In 1961, the MAAG significantly altered the course of the war by directing that advisors would be assigned to combat battalions, and that a Battalion Combat Advisory Team (BCAT) would consist of a captain, a lieutenant, and a senior non-commissioned officer. As U.S. combat support assets, such as helicopters and the U.S. Air Force aircraft, began to arrive in Vietnam, the influence and role of the advisor also increased. With all these indicators of an expanding U.S. commitment to the Vietnam War came another; U.S. casualties. The first airborne advisor to be killed was Captain Don J. York, who died on 14 June 1962 in an ambush of a vehicle convoy north of Saigon. Don York headed the Team 162 Killed In Action (KIA) list, which would eventually number 21 U.S. Army advisors, three Air Force Forward Air Controllers (FACs), and two Army advisors who still are counted as missing in action—the most of any division advisory detachment in the Republic of Vietnam."[78]

"By December 1965 the Brigade was upgraded to a Division, because two more infantry battalions (2nd Airborne Battalion and the 8th Airborne Battalion), along with an artillery battalion were activated. Concomitantly, advisors were assigned not only to combat, but also to combat support and combat service support elements within the Airborne Division. MACV Team 162 was the largest advisory effort of its kind. Advisors were actively engaged in all the Division's combat operations, training, and logistical upgrades. During this time the Airborne Division became the most aggressive in the Republic of Vietnam, participating in the critical battles of the war. The most notable werethe following: Hue and Phu Cat in 1966; Dac To in 1967; Khe Sanh, A Shau Valley, and the 'TET' Offensive in 1968; the defense of Tay Ninh in 1969;

Vietnamese Airborne Group shoulder sleeve insignia, used from 1955-1959. It features the white winged mythical horse "Pegasus" on a maroon background.

and the Cambodian operation in 1970. In addition, the Division had conducted three combat parachute jumps, one in Binh Long in 1965, and in Chuong Thien in 1966 and 1967."[79]

EARLY UNIFORM NOTES

When (MAAG) Military Assistance Advisory Group established an advisory detachment with the Vietnamese Airborne Group in 1955, these first MAAG advisors wore the abbreviated Army khaki uniform. Ribbons were not generally worn on the qualification badges. Airborne wings, CIBs, and such were worn on the left side. On the right side would be the advisor's nametag, along with the early South Vietnamese airborne group badges, which featured the mythical winged horse Pegasus in white on a maroon background. This was hung from the right pocket via a pocket hanger device, or pinned to the pocket. Headgear consisted of the khaki overseas cap with the early glides oval. The abbreviated uniform was banned in 1966 due to its French Colonial appearance, since it was felt it would conjure up hard feelings between the advisors and their Vietnamese counterparts.

By early 1960, U.S. advisor uniforms consisted of stateside fatigues with matching fatigue cap, to elaborate South Vietnamese airborne camouflage with red berets. Note arriving MAAG advisors wore the heavy-duty OG 107 cotton fatigues, shirt, and trousers made from cotton canteen. Full color insignia, such as a blue and white MAAG patch, were worn, including U.S. Army nametapes and basic qualification badges. The fatigue shirt was usually worn tucked in, and the trousers were worn with a black belt and the brass roller buckle. Black jump boots, highly spit shined, were worn bloused into the trousers. Fatigues headgear consisted of the spring up field cap, affectionately called the Ridgeway cap, made by the Louisville Cap Co., along with a privately purchased block spring up cap, or the 1951 cotton field cap, which was very popular. Unlike the "Spring Up" caps that could not be laundered, the cotton 1951 was much cooler and tended to last longer in the field. The 1951 field cap was usually worn with U.S. pin-back airborne wings and a set of the Vietnamese pin-back wings, and/or a combination of U.S. rank and parachute badges.

By 1962, most Airborne advisors were wearing the complete uniform of their Vietnamese host, to include the rose colored airborne beret. The camouflage uniforms that the Vietnamese Airborne wore consisted of a set of shirt and pants in what was known as the "bulletproof" pattern. This was a direct copy of the British SAS camouflage pattern known as the "wind proof" pattern. Worn above the left pocket would be the nametape, not a U.S. ARMY strip, done on the same camouflage pattern material as the uniform. No U.S. Army tape was worn at this time on the "bullet proof" pattern uniforms. Around 1963, Major Dan Baldwin lobbied and attained permission from MAAG for the American advisors to begin wearing the Vietnamese Airborne berets officially; unofficially, the Americans had been wearing them as early as 1960.

By 1965 the Vietnamese airborne had four camouflage patterns to choose from: the first, of course, being the "bulletproof" pattern; and the other three being descendants of the ARVN Ranger pattern, which was based on the early U.S. ERDL pattern. It is also important to note that prior to 1965, the Vietnamese Airborne wore the French Lizard pattern quite extensively, also.

4 SVN 9th Airborne Battalion rectangles. These were only worn on one side of the shoulder loop, and worn singly, never in pairs.

4 SVN 8th Airborne Battalion rectangles.

U.S. AIRBORNE ADVISOR, TEAM 162, VIETNAMESE AIRBORNE DET. TAN SON NHUT, 1964

"The Vietnamese Airborne was an Elite Force of all volunteers. These Airborne troops underwent 9 weeks of intense training, followed by a three week jump school, similar to the one at Ft. Benning, GA. These special volunteers received better pay, rations, quarters, and family benefits than common South Vietnamese soldiers because more was expected of them. The U.S. Advisors assigned to the Vietnamese Airborne were an elite group themselves, selected from the many airborne qualified NCOs and officers who volunteered to serve with the Vietnamese, and being selected to serve with the Su Doan Nhay Du was a great honor and privilege. Most of the advisors were young but experienced, with four extra months of training in Vietnamese language and customs and the MATA course taught at Ft. Bragg."[80]

Many of the U.S. airborne advisors had attended the MATA course (Military Assistance Training Advisory) at Ft. Bragg, NC, with part going to the army language school in Monterey, CA, for a seven week course, while some advisors went straight to Vietnam. "The MATA course contained seven courses; they were fa-

Photo of "Ridgeway" cap, favored by many of the early Airborne and Ranger advisors alike in the early 60s.

Upper torso shot of Capt. Sprague assigned to team 162, Vietnamese Airborne Det. Tan Son Nhut Airbase, Vietnam, 1964.

Upper torso shot shows the Airborne advisor's U.S. Army tape, name tape, "Jungle Expert" badge, pin-on Branch of Service (BOS) Infantry, and Captain Bars. Note the lack of a shoulder sleeve insignia on his left side, which should be the MACV patch.

Upper torso shot of Advisory Team 162 member shows the white "Master Blaster" wings, along with his civilian watch. Note the shoulder loops on his uniform.

Close-up of Vietnamese Airborne advisor provides a good view of his maroon Su Doan Nhay Du "Airborne" beret, "Jungle Expert" badge, and his 11th airborne Shoulder Sleeve Insignia showing a combat tour in the South Pacific. Note also the "Skillcraft" pen connected below the top button of his shirt.

miliarized with the old series weapons (WWII or Korea) that the Vietnamese were armed with at the time. They also covered map reading, radio procedure, etc. All of that was a waste of time for career infantrymen."[81] Throughout 1963, one battalion at a time was on standby.

This early Airborne advisor, Team 162, wears a set of the 1st model OG 107 cotton sateens with added shoulder loops. Full color U.S. ARMY tape, which is the machine made variation—not the silk woven style—is worn above his left pocket. Above the right pocket is a U.S. made white name tape. Below, on his right pocket is the U.S. Jungle Expert Badge, which features a Spanish Galleon, the Nina, Pinta, and Santa Maria, and The Caribbean Command. The patch could be worn on either pocket, but was usually worn on the right pocket, as seen here. Worn above his U.S. ARMY tape is a white set of Master Airborne wings. These are made on HBT material, and were originally on OG material, but constant washing and exposure to the hot tropical sun has bleached them white. Worn on both collar points are pin-on rank, (Captain) nd BOS (Infantry). This was not uncommon at this time. Note, no SSI is worn on his left sleeve, where his MACV or MAAG patch would be worn. On the opposite sleeve is the 11th Airborne, showing a combat tour during WWII in the South Pacific. Matching 1st model OG 107 cotton sateen trousers are worn with the black web belt and gold roller buckle. Headgear is the maroon Su Doan Nhay Du beret with the Gold bullion beret badge. Major Dan Baldwin was responsible for obtaining permission for the advisors of Team 162 to wear the red beret, and also to make combat jumps in or around 1961.

Typical WWII era M-1 helmet, worn by both the Vietnamese airborne and their advisors alike during operations and jumps. Note the single 1st aid dressing secured in the helmet net.

Early Su Doan Nhay Du wearing the early Airborne pattern camouflage and maroon beret with bullion Airborne beret badge.

Early photo of four Vietnamese in 1963-64. The man on the far left appears to be Airborne and wears the French lizard pattern jacket with pin-on airborne wings, OD cotton pants, black leather boots, and a maroon Airborne beret. Next to him, in the civilian clothes, is a friend who may be leaving for the Army soon. Next to him, a Vietnamese assigned to the 9th ARVN Infantry Division, as evident by the patch on his OG cotton fatigues. The Airborne soldier on the end wears a set of the Airborne camouflage, known as the "Bulletproofs," or the purple pattern, along with the early Airborne Brigade patch on his left sleeve, jump indicator, and tigerstripe boonie hat.

Captain assigned to the Vietnamese Airborne Brigade, Tan Son Nhut, 1964. Note the advisor wears his nametape on the left side of his Vietnamese airborne "Bulletproof" pattern shirt.

U.S. AIRBORNE ADVISOR, TEAM 162 5th VIETNAMESE AIRBORNE BATTALION, TAM HIEP, 1964

"These American advisors were an elite group, selected by the many U.S. Airborne sergeant and officers who volunteered to serve the Vietnamese Airborne Division. Most advisors were young, but not inexperienced. They were American sergeants and captains with four extra months in training in the Vietnamese language at the Ft. Bragg MATA Course. They were a bit different than the other American Paratroopers"[82]

"5th Vietnamese Airborne Battalion, along with this advisor, are based at Tam Hiep, a short distance from the Bien Hoa Air Base. The 5th was a crack outfit, which made a combat jump into Dien Bien Phu in November 1953."

Close-up of typical WWII era M-1 steel pot with helmet netting, like the one worn by the advisor during his practice jump (see top p. 88). Note the netting is much bigger on this one, and it lacks the 1st aid dressing. (Eric Halsey collection)

Black and white shot of Team 162 advisor after a recent practice jump. He wears the "Bulletproof" pattern, with a mixture of WWII and Korean War era gear, an M-1 steel pot, and the late model WWII jungle boots, which were produced too late in the war to see service. Note the M1A1 folding stock carbine.

Close-up showing the double pen pocket on the left sleeve.

Close-up shows the pocket with covered buttons of the "Bullet proofs" pattern camouflage shirt.

Close up of early silk woven Vietnamese Airborne Brigade patch, 1959-1965.

Close-up showing the button cuffs.

Close-up of the printed label, showing manufactured date as 04-1961.

Worn by this U.S. airborne advisor to the 5th Vietnamese Airborne Battalion is a pattern of camouflage that was known by many different names. It has been referred to as "Pinks," "Purples," the "British Windproof pattern," the SAS pattern, and the British "swept" pattern, as well as the "Bullet-proof" pattern by the Advisors who wore it. The pattern is a direct copy of the British Special Air Service pattern camouflage smocks worn in World War II. The ARVN Airborne adopted the pattern in the 1950s and it became their own. The Vietnamese clothing Production Center developed, printed, and produced this camouflage material after surplus stocks of the original material ran out in the early part of the 1960s. The French Colonial Forces in the first Indo-China war first brought this pattern to Vietnam. The shirt features a gas flap on the inside, shoulder reinforcements, and a large pen pocket on the left sleeve, popular among the Vietnamese officers. The early Co Vans to the Vietnamese wore only their name on the left side of the fatigues, usually done on a piece of matching material, and no U.S. Army strip. Above his name is a U.S. manufactured cloth "Master Blaster" airborne badge. On the right side is a Silk Vietnamese Airborne Qualification badge. Worn on the collar points are his B.O.S. (Infantry) and his rank; Captain. Early advisors like this one wore numerous styles of headgear. Many were made of the same material as the uniform, done in several different patterns.

Photo of two Vietnamese paratroopers. Both carry the PRC-25 radio and are armed with M-1 carbines. Note the lack of helmet covers on their M-1 helmets.

Portrait of SVN paratrooper taken in Saigon. Note the French/Vietnamese Fourragere on his right sleeve, 1964-65

Four South Vietnamese Airborne DI Rectangles, from top to bottom are: the 7th Airborne Battalion; 6th Airborne Battalion; 3rd Airborne Battalion; and the 7th again.

Four South Vietnamese Airborne DI Rectangles, from top to bottom are: the 11th Airborne Battalion (top two); the Support Battalion; and the 1st Battalion.

Early Su Doan Nhay Du beret, more of a cherry red without the bullion badge. Note the early ARVN Ranger pattern underneath the beret.

U.S. AIRBORNE ADVISOR,
TEAM 162,
TAN SON NHUT,
VIETNAMESE AIRBORNE,
1965

At this time the Vietnamese Airborne had six battalions: the 1st, 3rd, and 8th Battalions in the Saigon area; the 5th battalion at Tam Hiep (just off the Long Bien Highway before the turn to the road to Bien Hoa); 6th Battalion at Baria, just outside of Vung Tau; and the 7th battalion outside of Bien Hoa Air Base. This advisor is assigned to the Jump School at Tan Son Nhut and wears the Vietnamese ARVN Ranger pattern fatigues, modeled after the U.S. ERDL pattern. First intended for the Vietnamese Rangers, soon most elite units adopted this pattern, along with the "Bullet proof" pattern. The lightweight poplin shirt features two upper patch pockets, shoulder reinforcement, and shoulder loops. Worn on the left sleeve is the Vietnamese Airborne Brigade patch, which features a maroon square with a white parachute canopy, and a green bird. This patch was worn from 1959-1965. Worn above his right pocket is a U.S. made white name strip, taken from his OG 107 uniform. Worn on the left pocket is the winged St. Marcus badge, or the Jump Status badge. ARVN soldiers and their advisors on Jump status wore this pocket device, which features a yellow winged hand clutching a knife against a black background. This variation is a printed model. Helmet is an M1 steel pot with the M1C parachutist helmet liner, which is worn here unsecured

B&W shot of Team 162 Airborne advisor at Tan Son Nhut, 1965.

Additional shot showing the Team 162 advisor's left side. He wears the ARVN Ranger pattern with the 1959-1965 SSI at Tan Son Nhut Airbase in 1965.

Team 162 Airborne advisor, 1965. Note the outdated WWII era Tommy gun and the M-1 helmet with the Paratrooper liner, Tan Son Nhut, 1965.

Two Vietnamese Airborne SSI. The one on the left was used from 1959-1965. The other is the Vietnamese Airborne Division patch used from 1965-1975.

Right: *A very young SVN paratrooper wearing a set of the ARVN Ranger pattern and a maroon airborne beret with bullion style Airborne badge, 1965.*

Four 2nd Vietnamese Airborne Battalion DIs.

Four more Vietnamese Airborne Battalion DIs from the 6th Airborne Battalion.

Another young SVN paratrooper wearing a very light colored maroon Airborne beret with bullion style badge and a set of the ARVN Ranger camouflage that was also employed by the Airborne.

Vietnamese paratrooper wearing the early model Airborne pattern camouflage known as the SAS pattern, or "Bullet proofs," by the advisors. Also known by some as the "pinks" or "purples." He wears the maroon airborne beret with a hand sewn beret device, and wears the new SVN Airborne Division patch on his left sleeve, 1965.

SVN paratrooper wearing the airborne beret with what appears to be the bullion style badge. He also wears the ARVN Ranger pattern with the new Division patch. This is typical of many SVN photos to have color added afterwards, 1965-66.

and with no helmet cover. Web gear consists of M1948 cotton green suspenders connected to an M1944 pistol belt, an M1952 canteen cover with an aluminum canteen, a WWII era first aid/compass pouch, and two M1956 compass/first aid pouches. Underneath his ARVN Ranger pattern fatigues is worn a white T-shirt.

U.S. AIRBORNE ADVISOR,
TEAM 162,
TAN SON NHUT AIRBASE,
VIETNAMESE AIRBORNE,
1965-66.

Worn by this Team 162 advisor during a ceremony at Tan Son Nhut is the Army Tan Khaki uniform, consisting of cotton twill shirt and pants in khaki Army Shade #1. The long sleeve shirt was usually issued with the conventional trousers, but during 1959 regulations were amended to allow the wearing of the short sleeve shirt with the conventional trousers, which became very popular, and many resorted to cutting off the sleeves to make all shirts short sleeves. MACV designated the khaki and tan uniform as the normal duty uniform in Vietnam unless operating in the field, at which time fatigues were the uniform. "The Army khaki cotton

American Airborne Advisor from Team 162 wears the Khaki dress uniform for an award ceremony at Tan Son Nhut, 1965-1966.

Close-up of Su Doan Nhay Du beret with bullion style badge.

Close-up of miniature winged sword of St. Michael, patron saint and the protector of paratroopers. Sometimes worn as a beret badge, supposedly.

uniform, composed of khaki shade 1 cotton twill shirt and trousers, was the traditional mainstay of summer garrison duty attire. The Army color of khaki shade No. 1 was reserved for the hundred percent cotton twill of the uniform. The hue was a light beige brown color that was, as one better described it, 'a drab color that looked warm on a hot summer day.'[83]

"Cotton twill khaki clothing requires strict personal maintenance by each soldier. This was characterized by almost daily cleaning and starching of the uniform to present a suitable military appearance. As a result, the serviceman paid meticulous attention to the care and cleaning of his uniform. Garrison conditions enabled unit commanders to demand and expect this level of uniform perfection, which increased the soldier's pride in his personal appearance and his work."[84]

"The visible status of khaki clad troops was higher than those wearing the other uniforms for two reasons. Army cotton khaki was a seasonal duty uniform for all ranks, and most Army units were stationed in posts where summertime conditions prevailed."85 Trouser belt is the standard black cotton-webbing belt with a brass roller style buckle. Above his left pocket are basic parachute wings and a C.I.B. on his left pocket flap is an expert rifle badge. "The most prized award for the airborne advisor, as well as Infantryman, was the Combat Infantryman's Badge (C.I.B.). General Leslie J. is credited with the creation of the combat infantryman's badge.[86]

Worn on both shoulders of his khaki uniform are U.S. made E-5 chevrons, along with a pair of pin-back Vietnamese made sand cast wings. Airborne soldiers as well as Airborne Advisors wore private purchase Corcoran jump boots, highly spit shined, with the tan service uniform. The boots featured reinforced ankle supports and toecap, and both were spit shined more than the rest of the boot.

Worn on the advisor's head is the rose color Su Doan Nhay Du (Division Airborne) beret, which was worn by U.S. Army Advisors to the ARVN Airborne Division and its battalions. The Vietnamese berets were worn opposite of American berets, with the badge almost worn over the right ear in some cases.

U.S. AIRBORNE ADVISOR,
TEAM 162,
91st Company, 9th AIRBONE Battalion,
TET Offensive, Hue City,
1968

"The Vietnamese Airborne Forces served throughout the conflict as a national strategic reserve. By 1967, at strength, more than 13,000 paratroopers were formed into nine infantry battalions,

Airborne Advisor to the SVN 9th Airborne Battalion, Airborne Task Force, Quang Tri/Hue, "TET" Offensive, 1968.

B&W photo of Airborne Task Force advisor wearing his fatigue cap, which is very similar to the USMC caps.

Close up showing advisor's shirt. Note the name is where the Army strip should be. Note both the wings and CIB are hand sewn, while the jump indicator is a printed variation. Also note the button cuffs on the shirt's sleeves.

Five South Vietnamese Paratroopers take time to have their photo taken in Saigon. Note all but one wears the ERDL poplin uniform. The man in the center wears the early Ranger/airborne pattern, which is a very washed out pattern. All wear U.S. combat boots, 1967-68

three artillery battalions, and division troops battalion combat teams staged by parachute assault or helicopter insertion from Ton Sun Nhut air base in Saigon as reinforcements for two major battles of the war. Throughout the war, the parachute force was constantly committed to fight wherever the action was most dangerous."[87]

"The Airborne Task Force was comprised of the 2nd, 7th, and 9th parachute battalions. Several days before the 'TET Offensive,' the 9th Battalion was moved to Quang Tri City—several miles north of Hue. The 2nd and 7th Battalions were bivouacked at Post Kilometer (PK) 17, a Hamlet 17 km from Hue. At 2 am, the morning of TET, Capt. Mei Co, Commander of the 91st, shook his American advisor and said to get dressed, the 94th company was under an enemy assault and an attack on their position was imminent; the 91st and 94th companies were separated from the Battalion by several km. At 3 am the 9th Battalion came under attack from what was believed to be a company sized unit, but turned out to be a battalion sized force. By 4 am the 94th Co.,

Several very happy SVN Paratroopers wearing the ERDL Ranger pattern with the Divisional SSI and orange and gold French/Vietnamese Fourragere. Several even wear their ribbon bars—it is very likely that this photo was taken after an award ceremony, 1968-69.

Four different versions of the South Vietnamese Jump wings.

by attrition (wounded and dead), was down to 10 to 15 privates and a single NCO when they managed to withdraw back to the battalion headquarters; they had been overrun by a North Vietnamese Army regiment, and most of the para-troopers died fighting in place, advisor John church among them—an example of the airborne spirit, fighting to the last man! Daylight fared no better for the 91st company, which was under attack from a three battalion size force or larger. At 11 am the Ninth conducted a counterattack with two platoons, pushing the NVA back across some open rice paddies, where the Ninth held its position for three days until they were down to about a force of two platoons. Meanwhile, the 2nd and 7th Battalions had received orders to move to Hue. In route, they were ambushed by a regimental sized unit. The 7th Battalion was cut down to one company and one remaining airborne advisor, Captain Redding.

The 2nd Battalion, elements of the Vietnamese armored cavalry unit, and one company from the 7th Airborne Battalions, with Capt. Redding, entered Hue. Three days later, the battalion was ordered to conduct an airlift from Quang Tri City to Hue to support the 2nd and the 7th Battalions. After coordination, a Marine lift unit picked up the 9th Battalion and moved in to secure a section of the Citadel; that evening, the 91st Company made a night attack and seized a portion of a wall surrounding the city. Trying to capitalize on this and widen their position, the 2nd and 9th Battalions tried for several days—24 hours a day—to break through North Vietnamese defense. Later it was known that the area was defended by approximately two reinforced NVA battalions. The Airborne total strength, by this time (the 2nd, 7th, and 9th Battalions) hardly equaled one combat size battalion.

Later, the 91st Company was ordered to join the remaining elements of the 9th Battalion. The remaining elements of the 90th and, 91st, 92nd and 93rd companies went onto the attack while the 2nd Airborne Battalion attempted to secure some high ground—any gain in the wall at another part of the City. At one point in the attack the Americans assigned to the 91st called in air strikes using 250lb high drag bombs, but it still could not dislodge the NVA soldiers. The 7th Battalion was on reserve at this time, and days went by as the casualties became overwhelming, but true to the Airborne Spirit, there was no quarter asked or given.

Three 11th Vietnamese Airborne Battalion DIs and one 1st Battalion, which are all silk woven.

Four SVN Battalion DIs, from the top: the 9th; the 8th; and two from the 7th.

After two weeks the 9th Battalion was relieved by a fresh Task Force of reinforced Vietnamese Marine infantry, who two months later would finally secure the city of Hue."[88] "The 8th Battalion hit a N. Vietnamese Army Regiment head on to halt the enemy surge across the runway Tan Son Nhut and drive them off the air base. Meanwhile, two companies of the 8th Battalion were sent to assist the remnants of the 6th Battalion, now reduced to less than 200 men, where they defeated two enemy regiments at Dac To just three days earlier and stopped two North Vietnamese Army Battalions and the Joint General Staff compound. In Saigon, the 1st Battalion held off a threat on the Presidential Palace, cleaned out Cholon, and regained control of the radio station. The 5th Battalion returned to Saigon to assist in the fighting immediately after operations north of Da Nang, while the 3d Battalion was fighting NVA units at Go Vap and Xon Moi north of the capital. The 2nd Battalion and the 7th Battalion joined the 9th Battalion to break the siege in the Citadel. Previously, the 9th had fought 2 North Vietnamese Army Regiments in Quang Tri. Still undergoing training, the newly activated 11th Battalion repelled a reinforced VC Battalion attack on the Van Kiep Training Center."[89]

Vietnamese paratrooper wearing ERDL poplin uniform in the typical SVN cut. Note the pen pocket under his armpit. Note also the beret has been colored after the photo was developed. Circa 1968.

This Airborne advisor wears a Vietnamese style fatigue top, based on a U.S. utility shirt, using ERDL poplin material. Typical of most ARVN made garments, the shirt incorporates shoulder reinforcements and a small pen pocket under the arm, but lacks the traditional shoulder loops found on most Vietnamese made garments. Name strip is worn here on the left side with no U.S. Army tape. Above that are U.S. parachute wings and a Vietnamese made hand sewn C.I.B. On the left pocket is the Vietnamese Airborne Division "jump indicator," a cloth printed variation of the winged sword of St. Michael, patron Saint of the paratroopers. The Su Doan Nhay Du, or Airborne beret, is worn here, which is maroon in color. The badge is a bullion style variation, hand embroidered gold wire parachute and silver wire wings, along with a detailing of green thread. Airborne advisors carried a wide variety of sidearms as personal protection and as a sign of authority, the most common being the M1911 A1 pistol seen here in a black leather shoulder holster.

American Instructor gets ready to make a practice jump at the Vietnamese Jump School at Tan Son Nhut, 1968/69.

U.S. AIRBORNE ADVISOR, TEAM 162 "JUMP SCHOOL" TAN SON NHUT, 1969

"Relying solely on volunteers, the Airborne Division recruited approximately 4,000 men each year. Each volunteer underwent a special 9 week basic training course with an airborne cadre; he then went through a rigorous three week course of parachute training at the Airborne Division Training Center on Tan Son Nhut Air Base. The school was modeled after the U.S. Army jump school at Ft. Benning, Georgia. Classes were divided into three phases: ground week; tower week; and jump week. The students were awarded their parachute badge (jump wings) upon successful completion of the course, which included five qualifying jumps from a C-46, C-47, or C-119 aircraft. The school averaged 250 students per class."[90]

SVN paratroopers wait to make their last jump during Jump School, 1969.

"Army parachutists and the Vietnamese Airborne used the T-10 parachute assembly, which consisted of a backpack carried main parachute that deployed automatically by means of a static line; a parachute harness with a saddle and straps around the chest, legs, and back; and a reserve parachute, hand-activated with a rip cord handle. During 1962, following a series of mishaps caused by wind gusts dragging paratroopers across the ground, a quick release system was added to the assembly. This allowed soldiers a rapid means of disconnecting the main canopy upon landing."[91]

"Paratroopers jumped with individual equipment and sleeping bags inside the aviator's kit bag, secured by H-harness be-

Rear view provides a good look at the advisor's outdated WWII T-10 parachute.

Full view of Airborne Instructor with T-10 parachute and lightweight rucksack.

Vietnamese paratrooper wearing the ERDL poplin uniform in the traditional Vietnamese cut.

neath the reserve parachute. During 1963 the T-10 parachute harness was fitted with quick release assemblies on each shoulder, which allowed the jumper to disengage the canopy upon landing."[92]

The Airborne advisor wears the M1 steel helmet fitted with the M1-C Parachutists Liner. "The Parachutist's liner differed from the standard infantry version in having an adjustable web chin cup. This buckled to A-straps on each side of the liner and could be roughly tightened to fit. Designed to be buckled during a jump to prevent the helmet being lost or causing injury. The chin cup was typically worn loose or removed altogether during ground operations."[93]

Team 162 advisor wears a set of the U.S. ERDL rip-stop fatigues with a Vietnamese Airborne Division patch worn on the left sleeve. Note the ERDL trousers are rolled up; prior to boarding the plane, the trousers would be tucked in and secured inside the advisor's jump boots. In his hand is his M-16 rifle, which would be secured along with the PRC/77 radio in the "(WEIC), the Weapons and Individual Equipment Carrier. The WEIC was used to secure weapons, radios, and crew served weapons components, along with ammunition. Up to 100 lbs of equipment could be jumped in this way secured inside the adjustable envelope. The WEIC was hung from the D-rings on the harness and lowered on its 18 ft rope just prior to landing."[94]

Four Vietnamese Airborne Battalion DIs: 11th; the Engineer Battalion; and two from the 1st Battalion, bottom row.

Four Battalion DIs from the 1st Airborne Battalion.

U.S. AIRBORNE ADVISOR, TEAM 162, 11TH AIRBORNE BATTALION, LAM SON 719 1971

"Lam SOM 719, the attack into Laos in February 1971." For the first time, a MACV order prohibited advisors from accompanying their units into combat. The purpose of Lam Son 719 was to cut the Ho Chi Mihn trail southwest of Khe Sahn, in Laos, near where the trail crossed Highway QL9. Large supply dumps in the area were also to be captured and destroyed. The 1st ARVN Infantry Division, VN Marine Division, and the Vietnamese Airborne Division were all participating. The Red Hats, with an attached armor brigade, moved quickly into Laos during the early days of the operation and set up a series of fire bases, which were to be used to support other units as they passed through, going west to cut the Ho Chi Minh Trail."[95]

"Command concern over U.S. casualties and the growing reliance on 'Vietnamization' of the war prompted the directive. An advisory base was established at Khe Sanh when the units crossed into Laos. While individual airborne battalions fought the North Vietnamese valiantly, Vietnamese commanders were unable to

Three Airborne DIs from the 1st and 2nd Battalions.

U.S. Airborne advisor to the Vietnamese 11th Battalion, 1970.

Close-up of Vietnamese Airborne hat, very similar to the USMC fatigue cap. This one is in the brown dominant ERDL pattern, with silk woven airborne wings sewn to the front.

Inside of airborne hat shows a peach colored liner with the maker's tag and size.

synchronize U.S. fire support assets. Heretofore, advisors had accomplished this, and now their absence created a void in paratroopers' ability to orchestrate the battle. During the fighting on the way to Tchepone, five of nine battalion commanders were killed or wounded, and when Firebase 31 fell, the North Vietnamese Army captured the 3rd Brigade commander and his staff. The Airborne Division's withdrawal from Laos resulted in a long stand down period while the leadership attempted to fill its depleted ranks and began to assume an even greater responsibility for prosecuting the war. Notwithstanding the deficiencies that emerged during LAM SON 719, U.S. troop withdrawals continued unabated. President Nixon was determined to decrease U.S. involvement in Vietnam. The drawdown affected not only U.S. units, but also had an impact on the advisory effort. Although Vietnamization was in full swing, Headquarters MACV issued strength reductions for all detachments. By the fall of 1971, advisors had been removed from infantry battalions in all ARVN divisions except those in the Airborne Division. However, Team 162 was not totally protected from the cuts. Advisory positions in the artillery, engineers, and combat service support units were either eliminated or reduced to one individual. But, the decision to protect the BCATs in the Airborne Division proved to be decisive when the NVA launched their long anticipated offence in 1972."[96]

Close-up of bullion style airborne beret badge.

SVN paratrooper wearing the ERDL, U.S. boots, and the maroon Airborne beret, 1969

Close up of paratrooper with longer than average hair. He wears the early airborne pattern fatigues and the maroon Vietnamese Airborne beret.

SVN paratrooper wearing the ERDL pattern in rip stop. Note the strip on his left side E-54 boots—they appear to be U.S. Jump boots.

"The Easter offence, which began on 31 March 1972, introduced a new face of war inside Vietnam. North Vietnamese Defense Minister Giap directed conventional attacks in Military Regions 1, 2, and 3, involving the commitment of practically all his regular forces. These division size formations, well balanced in armor, infantry, and artillery, were oriented on the destruction of the ARVN, trapping if possible the remaining U.S. personnel in the country. The basis of North Vietnam's action revolved around the assumption that Vietnamization was a failure and that the U.S. public was so adverse to continued involvement in the Vietnam War that President Nixon would be unable to bolster the Thieu government. However, North Vietnam had grossly underestimated the U.S. response.

Giap's offensive stretched the Airborne Division even more than LAN SON 719 had. During the initial stages, brigades were

Right: *Close-up showing advisor's ERDL ripstop shirt with U.S. Army strip, subdued wings, silk woven Vietnamese Airborne wings, and a silk woven Vietnamese Airborne Division Shoulder Sleeve Insignia (SSI). Note the lack of either U.S. Army rank on his uniform or the jump indicator badge. This goes to show that not all uniforms were "flashy," and some were missing certain items. This one appears to be a replacement, in which the advisor did not have enough time to have everything put on at the tailor shop.*

Full figure view of American advisor to the Vietnamese Airborne Battalion, Team 162 Vietnam, "LAM SON 719," 1971.

Full figure view of American advisor to the Su Doan Nhay Du, after the "LAM SON" campaign.

deployed along Rocket Ridge in Military Region II and Quang Tri Province, and to reinforce the defense of An Loc. Fighting was intense, and casualties were very heavy. Because of strategic location north of Saigon, the NVA strangle hold on An Loc caught the attention of the international media. The 1st Brigade helped break that siege and stave off what could have been a disasterous defeat. By late June, the NVA stopped at An Loc, and the Airborne Division was completely assembled North of Hue and marshaled for an attempt to wrest Quang Tri City from the North Vietnamese occupation. The fact that it was the only province capital in North Vietnamese hands gave it an importance far beyond its tactical significance. By frontally attacking this fortified position, both the Airborne Division and the Vietnamese Marines frittered away the strength of their units. The Airborne battalion's advisors earned their pay by coordinating mass amounts of U.S. assets that were sent from the U.S. to stem the invasion. While U.S. firepower, practically U.S. tactical air support and B-52s, provided the margin of victory, it could not replace the experi-

Right: *SVN Paratrooper wearing a four-pocket shirt in the ARVN Ranger pattern. Both his shirt and matching trousers are styled after the U.S. Jungle fatigues. These appeared late in the war and are very hard to find with any insignia still on them.*

Two SVN paratroopers pose next to several M16 rifles. Note the number 99 painted on their M-1 helmets. Both wear ERDL fatigues. The one standing has a strip that reads V-100.

SVN Paratrooper wearing a two-pocket ripstop ERDL pattern shirt and trousers. Note the shirt features two pockets similar to the ones on the U.S. jungle fatigues, but lacks the two bottom pockets.

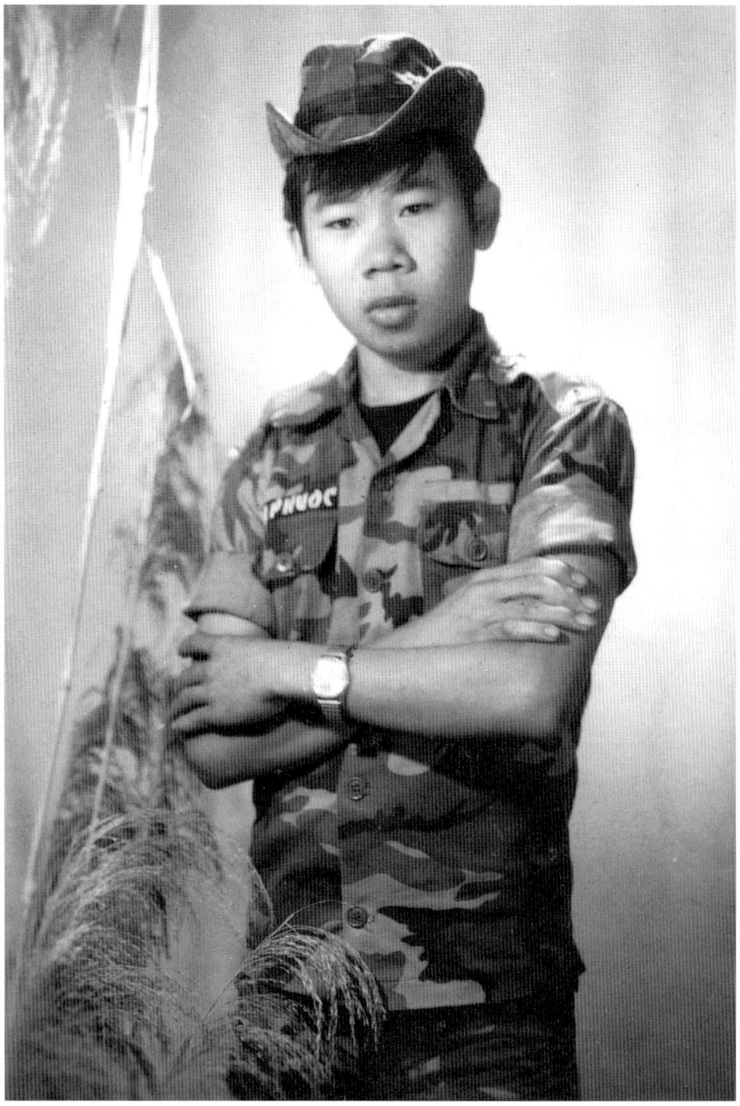

SVN Paratrooper wearing the ARVN Ranger pattern fatigues with matching boonie hat.

ence and elan that was lost on the battlefields of Quang Tri, An Loc, and Kontum."[97]

The uniform worn here is the late war ARVN camouflage issued to the Airborne Forces and many of the Airborne advisors. It was produced with U.S. lightweight cotton poplin rip stop material. The shirt resembles a U.S. utility uniform, and is printed with the Highland (Brown Dominant) ERDL pattern material. The shirt features shoulder loops and reinforced shoulders typical of Vietnamese made garments. On the left sleeve is a silk woven ARVN Airborne Division patch, along with a U.S. Army Ranger tab. Twin bars of a Capt. and the (BOS) crossed rifles of the Infantry are worn on the collar points. Name and U.S. Army tapes are worn above the two chest patch pockets. Above his U.S. Army strip is a U.S. produced set of "Master Blaster" airborne wings. On the front of his shirt, between the top buttons is the ARVN equivalent of his rank; Dai Uy. Once again the Su Doan Nhay Du beret is worn, Vietnamese style.

4

U.S. Ranger Advisors to the ARVN Rangers (BDQ) 1960-1971

In 1960, the Biet Dong Quan were created as a counter-guerilla force to track the cunning Viet Cong and strike at them in their jungle sanctuaries. These units were also intended as an elite back up force for districts in trouble.

In May of 1960, Col. William Ewald's Special Forces Training Teams arrived in Vietnam to select the sites for three training centers and begin training by June. These first three centers were located at Da Nang, Nha Trang, and Song Mao. The training course at Nha Trang would become the South Vietnamese Ranger Training Center at Duc My, and the training responsibilities switched from Special Forces control to the Regular Army advisory teams.

South Vietnamese Ranger wearing ARVN Ranger beret and tan service uniform similar to the Army's Tan Class B uniform. Circa 1964.

Vietnamese Ranger companies were trained and then distributed throughout the countryside. It was about this time that President Kennedy wanted to increase aid to Diem without the commitment of U.S. combat troops. The Government of Vietnam wanted a special type of U.S. Advisor, who could join the Rangers in the field and fight side by side with the troops, but also train them in guerrilla warfare and counterinsurgency.

These American Advisors, along with the cream of the crop Vietnamese Cadre, would take those newly created Ranger units through the counter guerilla course, along with a two week field operation against live pop-up targets in Indian Country.

Before receiving orders for Vietnam and assignment to a Ranger unit, the majority of officers were required to attend the (MATA) Military Assistance Training and Advisory course at Ft. Bragg, which included courses on Vietnamese customs, language, and duties of a U.S. Advisor.

Advisory Teams consisted of a Team Leader, a captain, an ordnance specialist who was an Sp5 or Sp6, and a signal specialist of the same rank. By May of 1962, the Spc were replaced by Infantry M.O.S. trained non-commissioned officers.

During 1965 and 1966, Rangers were assigned to ARVN divisions and suffered countless and unnecessary casualties as a result.

The Vietnamese Rangers, along with their U.S. advisors, fought in such famous battles as Dong Xoai, Plei Me, Co Do, Dak To, Ben Het, and Kontum.

1st SPECIAL FORCES GROUP (AIRBORNE) SFGA
ARVN Ranger Advisor
Da Nang,
1961-1962

Considering the U.S. Army had done away with Ranger formations after the end of the Korean War, the Green Berets were also tasked with ranger training requirements in foreign countries. One of the constant problems that blurred the line between unconventional Special Forces doctrine and their actual field practice was using Special Forces as ordinary military instructors. The Green Berets were directed to train sixty ARVN Ranger Companies at three training sites. These training facilities were actually run by 30 instructors from the 7th Special Forces Group Airborne. Early in 1961, these personnel were replaced by a nine-man Mobile Training Team, with four personnel from the 1st SFGA and five

Full front view of early 1st Group, Special Forces advisor to the newly created ARVN Rangers, 1961.

Full figure view with emphasis on the advisor's right side provides a good look at the advisor's WWII era double buckle boots. Note the Green Beret is carried out of view in his back pocket, and is not worn on operations.

Full view of early 1st Special Forces Group, Airborne advisor with emphasis on the advisor's left side provides a good look at the minimal gear worn by the early advisor.

soldiers from the 25th Infantry Division. The four Green Berets were deployed to Da Nang and Nha Trang to oversee ranger cadre development in Ranger tactics and techniques, but not at the unit level. It was training for selected individuals prior to this team's arrival. This early Special Forces Group arrived in Vietnam in November 1961 with eight people (four officers and four enlisted), and at this time all training had been consolidated at one site—Trung Lap, and already had the maroon berets and the BDQ patch. The selection process of who were to become "Rangers" at that time was rather unique. The ARVN infantry battalions in the early days consisted of three rifle companies. President Diem would select the best company from each battalion, and they would be designated Rangers. These new Rangers would be taken through the "blitz" counter-guerilla course that would last two weeks in the field operating against live enemy.

The advisors learned through experience that by using one company as a sweeping force and another company in a blocking force, the Rangers' KIA and VC prisoner counts doubled. Unfortunately, the best units would be stationed in Saigon as Diem's own palace guard, not in the field where they were needed.

When the first U.S. advisors arrived in Vietnam to establish the Ranger training sites, they wore the uniforms of stateside troops. These consisted of the heavy cotton OG 107 utility shirt and trousers. The fatigue shirt was single breasted with a collar, incorporated two patch pockets with a single button flap, and plain sleeves that could be rolled up when permitted. The fatigue top, along with a white T-shirt, was usually worn tucked into the trousers, unless determined by higher ups that it could be worn outside if it was considered hot. Matching trousers were equally uncomfortable; they featured two front and two hip pockets with buttoned flaps, button front closure, and button take-up tabs on the waist. The trousers were worn either tucked into the black boots or rolled up.

Headgear at the training centers consisted of the Ridgeway cap with white engineer tape, with the word "instructor" printed on the front. Pinned to this and above would be the instructor's

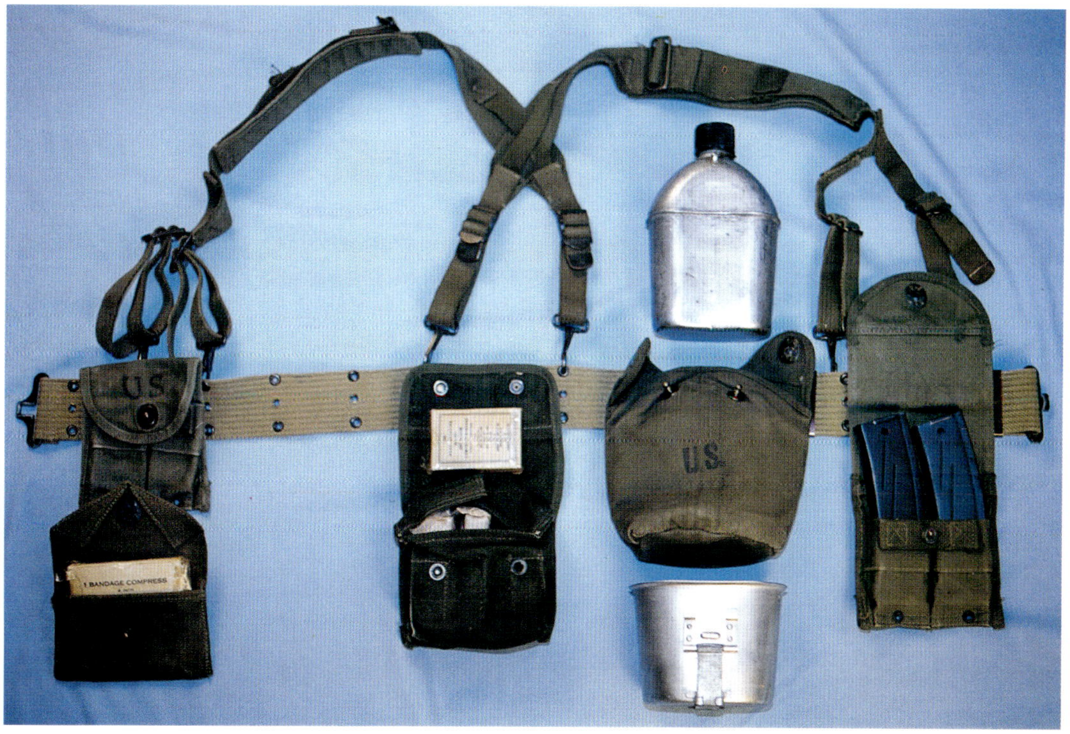

Layout of typical ARVN Ranger advisor's LBE, including WWII Marine Jungle first aid kit, metal canteen, and compass/first aid pouch. Note also the Korean War era magazine pouch for the long M-1 banana magazines.

rank, along with a set of U.S. pin-back airborne wings. The Ridgeway cap, named after Gen. Ridgeway, came into service about 1953 when it was decided that the 1951 patrol cap did not give soldiers a professional appearance. These caps, produced by the Louisville Cap Co. of Louisville, Kentucky, were guaranteed to spring back, look sharp, and not wrinkle or fade. By the late fifties and early sixties, most troops were wearing these caps. This 1st S.F.G.A. officer wears a fatigue shirt cut from a Korean War era shelter half, and styled after the first pattern OG 107 shirt. This material was water repellent and mildew resistant longer than the cotton fatigues. The material was printed with a disruptive camouflage pattern using selected organic dyes. It is a duotone reversible material with a leaf pattern in green colors for spring and summer and a brown color for fall or winter operations. This pattern was known as either the "Mitchell Pattern," or "Maple Leaf" pattern. Many early advisors liked the Mitchell pattern's

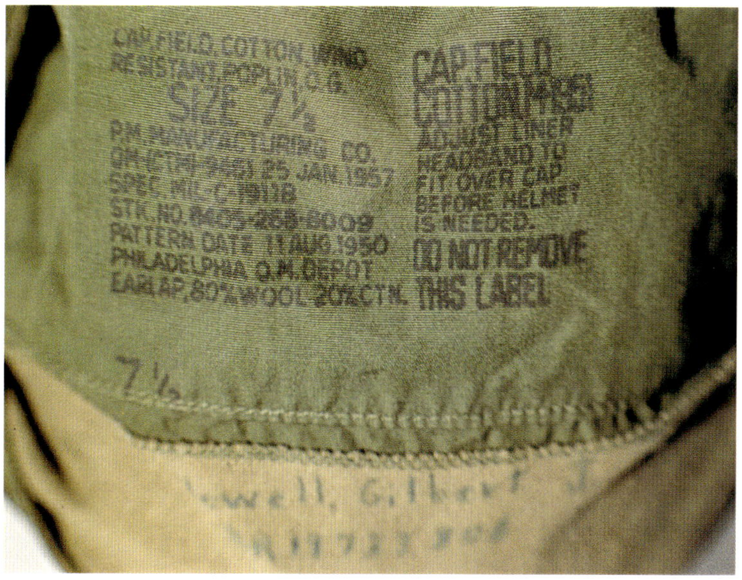

Close up of printed label inside advisor's Cap, Field, Cotton, and M1951.

camouflage capabilities, and Vietnamese garments were often too small for the average Americans. Underneath, he wears a standard G.I. cotton white T-shirt.

Russet brown WWII combat boots, known as the "double buckle boots," came into service in 1943 and are still being worn here in 1961. The boots had an all leather lower with a canvas/leather upper and a buckled cuff. These boots were better than their all leather counterpart, but because of the stitching, the sole often separated from the boot.

This "Green Beret" wears the Cap, Field, Cotton, M1951, better known as the "Patrol Cap," instead of his signature headgear, which just this year has become the primary symbol of Special Forces. First authorized in 1961 by President Kennedy, the wool beret in Army Shade 297 was followed by the adoption of a gold flash for the 1st Group in October 1961.

Web gear consists of: M1948 suspenders supporting an M1936 pistol belt, along with a single Universal pouch; a WWII first aid pouch; and a single aluminum canteen in its cover. The weapon is the full auto M-2 Carbine. Note the single "pineapple" grenade on his suspenders.

Field Command Advisory Detachment, Senior Advisor. 20th Battalion Kontum, 1962

In 1962 the American Advisory Teams consisted of a Team Leader who was a Captain, an Ordinance Specialist who was a Spc. 5 or Spc. 6, and a Signal Specialist of the same rank. The Specialists were later replaced with Infantry Non-commissioned Officers, and Sergeants with the Infantry (MOS) by the recommendation of the then Senior Advisor Major Daniel I. Baldwin III. The Senior Advisor answered directly to his boss, General Duong Van Mihn ("Big Mihn"), who commanded the III Corps, who wanted All-American Advisors inspecting the Ranger training centers at

Full view of early Ranger Advisor to the Field Command Advisory Detachment, Kontum, 1962.

Full figure view provides a good look at the Ranger advisor's M1A1 folding stock carbine rifle and PRC-6 "handie-talkie" radio, operating outside of Kontum, 1962.

Trung Lap and Duc My, along with the training of the three Special Battalions; 10th, 20th, and 30th. The 20th Battalion, like the 10th and 30th, were just starting to get their feet wet. In Sept of 1962, American advisors began leading companies of ARVN Rangers on extended operations, and on separate missions. At this time a company consisted of roughly 96 men, mostly ethnic Montagnards, Nungs, Cambodians, and Vietnamese, who, despite their ethnic backgrounds, worked well together and provided a unit of cohesiveness. The extended operation was to last 30 days+. Every five to six days the company would be re-supplied by air with food and ammo. This was a unique part of the operation. The advisor or senior ARVN Ranger would send up an orange balloon, which was about a meter and a half in size, and the pilots flying the re-supply plane would drop it on the balloon, which worked quite well.

"The 20th Battalion came across several Jarai tribe villages in which a marauding band of VC had been raping, murdering, and stealing rice. The company stayed in the village for several days helping the wounded and feeding the hungry before setting out after the VC. They tracked them all the way to the Laotian border, where the Vietnamese commander, Dai Uy Chein, wanted to follow into Laos. The American, Major Baldwin, informed him that they were not allowed to cross the border, but to no avail. The next day scouts ahead of the main element smelled the smoke and cooking fires of the VC. The Company was broken down into assault teams and support teams and crawled into position to attack the VC camp at first light. The attack was successful, with 85 VC dead and only two escaping. This was the first time the M79 grenade launcher was used in combat by an American, who killed two VC who were fleeing during the melee. After the AO was mopped up and all enemy weapons and other items of intelligence value were gathered, the company was prepared to move out when a lone sniper shot and mortally wounded the Vietnamese Commander, Dai Uy Chein. The company was still five to six days away from Kontum, and had no radio contact at all. Dai Uy Chein died the next day during movement back to Kontum."[98]

This early advisor typifies men assigned to the Field Command Advisory Det. in Kontum in 1962. The advisor wears a pattern of camouflage commonly called the leopard pattern, or "Duck Hunter." Copied after the World War II U.S.M.C. pattern,

Full view with emphasis on the advisor's left side provides a good look at his cut down bayonet on his web belt and WWII era "pineapple" grenade.

Close-up showing the advisor's cut down bayonet in its sheath and WWII era compass/first aid pouch on his M1936 pistol belt.

early U.S. personnel purchased these fatigues at such sporting good outlets as Sears and Roebuck. Both the jacket and pants incorporate the M1941 "Burst of Glory," or thirteen star metal tack buttons. His pants are also modeled after WWII fatigues, having two upper hip pockets instead of the lower thigh pockets. They also incorporate button take-up tabs on the waist.

His matching cowboy hat, although slightly darker, is a Vietnamese copy of the early French design, and was very popular with both the American Advisors and the Vietnamese Rangers.

Field gear worn here is a combination of WWII and Korean era items. The M1936 pistol belt is being supported by the M1948 cotton X-suspenders, along with two Mk2 hand grenades. The advisor carries the PRC-6 Radio. This small, hand-carried Korean War radio set resembles a hand telephone. It provided two way radio communications over a distance of about a mile. The radio also came with an external handset. His rifle is the M1A1 folding stock Carbine, popular among advisors. He carries his .45 in the russet brown World War II holster. On his pistol belt is a cut-down M-5 bayonet with a solid plastic handle. During the 50s and 60s, hundreds of M-5 and M-6 bayonets were modified by cutting off the O ring and removing the hilt, turning them into fighting knives. These were then distributed to countries receiving military aid, such as Vietnam.

Close-up of cut down bayonet with black M8 scabbard.

U.S. RANGER ADVISOR, 43RD BDQ ARVN RANGERS, VINH LONG/SADEC, 1964

In 1964, this reconstruction of Captain Rau was assigned to the newly-formed 43d ARVN Ranger battalion, BDQ, which was undergoing training at Trung Lap Ranger Training Center. The Captain arrived at the conclusion of individual training and joined the process during squad and platoon level operations. In a very short period company and Batt. Operations were concluded, and the 43d BDQ moved was based in the Vinh long/SaDec area in the ninth ARVN division's area of operations. Since the 42nd and 44th battalions were undergoing training concurrently at Trung Lap, immediately after the 43d battalion, the 4th Corps, ARVN used the 43d Rangers as a "fire brigade." In the next 30 days operations were conducted from Ap Bac, in the "Plain of Reeds" on the Northwest to Ben Tre on the Northeast, Ha Tien on the South China sea/Cambodian border on the Southwest, and Ca Mau and the "U. Mihn forest" on the Southeast. A long vehicle convoy to the seven mountain area, air assaults in every province, 4th Corps amphibious landings off the Bassac, and Co Chien branches of the Mekong River became commonplace to the 43d BDQ Rangers. Towards the end of June the battalion settled into a routine of operating under OPCON to the 13th, 14th, or 15th regiment of 9th Division, with most operations involving a series of VC contacts being conducted in Vinh Binh Province. June, July, and August were filled with one operation after another being typified by a "meeting" engagement leading to positive violent execution and successful conclusion to battles. Casualties mounted, and replacements streamed in during the summer of 1964. However, the loss of 3 company commanders in one battle handicapped the 43d's morale for a short time. In the first weeks of September political unrest touched the 43d BDQ as the Commanding General of 4th Corps, element of the 7th ARVN Army of the Republic of Vietnam division with attached units marched on Saigon. The 43d BDQ was spearheading this coup d'état. With advisors in tow, Sgt. First-Class John McCoy had joined Capt. Rau in June to complete the two-man advisory team. But the two left the coup forces in TAN AN and drove ahead to warn General Westmoreland of the impending coup. Things quieted down in the next several days, and Capt. Rau was reassigned after the 43d BDQ returned to Vinh Long. SFC McCoy remained with the battalion to lend his experience in three ways to the new four-man advisory team, headed by a Capt. David Fried. On September 26, 1964, two weeks after the aborted coup, on the 43d BDQ's next operation, SFC John McCoy was K.I.A.[99] The textile research conducted during the Korean War led to the development of more durable cotton material with the same weight as the herringbone twill fabric. During 1952, cotton sateen fabric was manufactured in Olive green shade 107 and produced the new utility jacket and trousers.[100] The U.S. Army used the official designation "utility"

Full figure view of ARVN ranger advisor to the 20th Special Battalion, 1963.

Full shot with emphasis on the advisor's front and right side. Note the captured VC flag in his left hand.

Close-up provides a good look at the advisor's pin-on jump wings, CIB, and early OG 107 cotton sateens, which have been modified to short sleeves. Note also the twin cell magazine pouch on the stock of his M-1 carbine.

Close-up of white WWII era tag for his Cap, Field, OD with visor.

Close up of early ARVN Ranger beret minus the badge. Note the tag and the huge "cocks comb" of the beret.

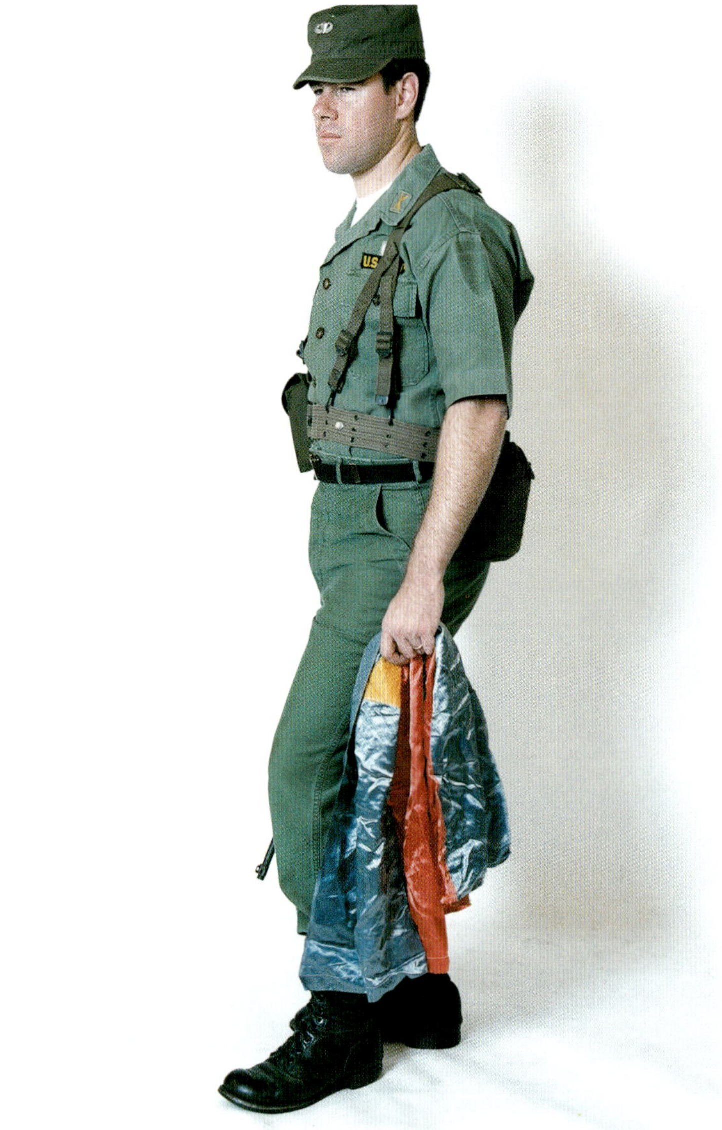

Full figure with emphasis on the advisor's left side shows the minimum amount of web gear worn on some operations with the 20th Special Battalion.

Full figure of advisor showing off captured "war booty," a VC flag. Note he holds it upside down, to show disrespect for the communists.

for this uniform, which was consistent with the terminology of other services. The utility, or "fatigues" were intended to be loose fitting, although the shirt was normally worn tucked into the trousers and the trouser bloused when wearing combat boots. Army soldiers continued to call these fatigues because of the fatiguing duty performed while wearing them.[101] The first model OG 107 sateen shirt worn here came with flat, brown buttons. Note, the shirt has been modified to a short sleeve garment. The matching trousers also featured the flat brown buttons, featured two rear pockets and two front pockets, and were worn with a black web trouser belt with a brass roller buckle.

Above his left pocket is the "U.S. Army" distinctive tape, first authorized 27 Oct. 1953. The woven label has golden yellow block lettering on a black background. A nametape was added by local unit directives above the right pocket and contained the soldier's last name, first initial. Early name tapes were often in branch colors, but most were made of white engineer tape with black lettering.[102] Above the U.S. ARMY strip is a pair of basic Airborne wings. On the left sleeve he wears no Shoulder Sleeve Insignia (SSI), but if he did, it would be the MACV SSI. Worn on his right pocket is the "Jungle Expert" school patch, showing he is a graduate of the Jungle Warfare Training School at Fort Sherman, in the Panama Canal Zone. The patch consisted of the Caribbean command patch, with the addition of an arc that read "jungle expert." Worn on his right collar is a Vietnamese-made hand-sewn 0-3, Captain's bar, along with U.S.-produced cross rifles (Infantry) on his left collar point. Headgear consists of another example of the M1951 field cap, or "patrol cap." This one has the officer's airborne wings pinned to them. Usually, his rank would be affixed underneath these wings, but this officer has chosen to wear only his wings.

Army web gear is shown here in typical fashion, and consists of a green M1944 web belt, supported by the M1948 cotton suspenders, a single Korean War era magazine pouch, which would accommodate 4 of the larger "banana style" magazines for his M-1 carbine, along with two aluminum canteens and covers. Note the twin cell magazine pouch secured to the buttstock of his carbine, which holds two of the smaller magazines. Footwear is a pair of black leather combat boots that resemble the jump boot, but lack the ankle reinforcing and the slant heel. In his hand he carries a captured VC flag.

Sgt. First-Class David Williams, 42nd BDQ, in 1964-1965. He wears OG 107 cotton sateens with the yellow Vietnamese Fourragere, pin-on CIB, and maroon ARVN Ranger beret. This photo was taken in a small studio and later distributed throughout the Delta as a wanted poster by the Viet Cong. The Vietnamese Rangers discovered this and burnt the small studio to the ground. It is unknown if the VC photographer was inside the studio when it burned to the ground. (Dave Williams collection)

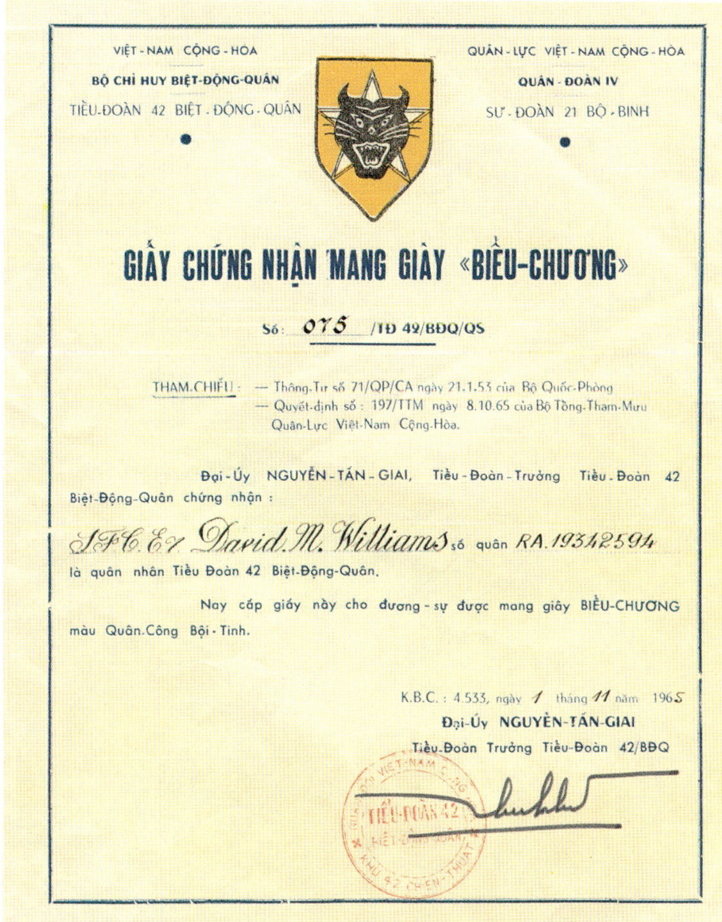

VIỆT-NAM CỘNG-HÒA
BỘ CHỈ HUY BIỆT-ĐỘNG-QUÂN
TIỂU-ĐOÀN 42 BIỆT-ĐỘNG-QUÂN

QUÂN-LỰC VIỆT-NAM CỘNG-HÒA
QUÂN-ĐOÀN IV
SƯ-ĐOÀN 21 BỘ-BINH

GIẤY CHỨNG NHẬN MANG GIÂY «BIỂU-CHƯƠNG»

Số: 075 /TĐ 42/BĐQ/QS

THAM-CHIẾU: — Thông-Tư số 71/QP/CA ngày 21.1.53 của Bộ Quốc-Phòng
— Quyết-định số: 197/TTM ngày 8.10.65 của Bộ Tổng-Tham-Mưu Quân-Lực Việt-Nam Cộng-Hòa.

Đại-Úy NGUYỄN-TẤN-GIAI, Tiểu-Đoàn-Trưởng Tiểu-Đoàn 42 Biệt-Động-Quân chứng nhận:

SFC E7 David M. Williams số quân RA.19342594 là quân nhân Tiểu Đoàn 42 Biệt-Động-Quân.

Nay cấp giấy này cho đương-sự được mang giây BIỂU-CHƯƠNG màu Quân-Công Bội-Tinh.

K.B.C.: 4.533, ngày 1 tháng 11 năm 1965
Đại-Úy NGUYỄN-TẤN-GIAI
Tiểu-Đoàn Trưởng Tiểu-Đoàn 42/BĐQ

SFC Williams' BDQ Award.

U.S. RANGER ADVISOR, TEAM 51 42ND ARVN RANGERS, AN XUYEN PROVINCE, VIETNAM 1964

Team 51 was stationed at Bac Lieu, in South Vietnam's Delta region, and was home of the fierce 42 ARVN Rangers. The Americans who made up team 51 were Captain Robert Butler, commander, the XO Lt. Dale Shipley, and SFC David Williams. Unlike other advisor teams, all three members accompanied the battalion on operations in the An Xuten Province. Team 51 was a cohesive unit that worked well with each other and their Vietnamese counterparts.

An almost infinite variety of military hats, ranging from camouflage to OD cowboy hats were worn by U.S. Army personnel

Right: *Captain Robert W. Butler on operations in the Delta region of Ca Mau with the 42nd Vietnamese Rangers. He wears first model jungle fatigues with full-color U.S. Army in name tapes. Around his neck is the HQ red/yellow scarf. Note the Branch Of Service and his rank are both pin-on. Note also the lack of a rifle, and his locally purchased cowboy hat. In is right hand is a map of the area. (Dave Williams collection)*

ARVN Ranger commander Major Nguyen Van Biet, wearing Vietnamese Duck Hunter camouflage fatigues and smoking the cigar, shows off captured Viet Cong weapons, 1964. (David Williams collection)

Sgt. First Class David Williams with radio operator during operations in Ca Mau. Note the sam pans behind them on the canal, which were used earlier to get them to their objective. Sgt. Williams wears the first model jungle fatigues, early jungle boots, and a locally produced cowboy hat. Over his shoulder is a M2 fully automatic carbine.

Upper torso shot provides a good look at the Lieutenant's Vietnamese rank and pin-on BOS.

Upper torso shot provides a good look at the Ranger advisor's tan cowboy hat and red/yellow scarf from the HQ Company.

Full front view of 1st Lieutenant, Team 51, 42nd ARVN Rangers, Advisor compound, Bac Lieu, 1964.

Full figure view providing a good look at the advisor's second model jungle boots and his side-arm in the black leather holster.

Full view of Trung Uy. Note the HQ red/yellow scarf.

in Vietnam. Many hats had only local authorization at best. For instance, numerous locally made "bush hats," or "cowboy," as seen here, proliferated within advisory detachments and other training elements. Some of the cowboy-style hats became a hallmark of Vietnam advisory duty from 1962 to 1966, but interest in them began to wane once the eminently practical and individualistic tropical hat became widely available.[103] This led one advisor, LTC Robert Parrish, Recondo advisor, to refer to the Vietnam War as the "Hat War." This advisor wears another example of a locally-purchased tan cowboy hat, influenced greatly by both the early French and American movie heroes, like John Wayne.

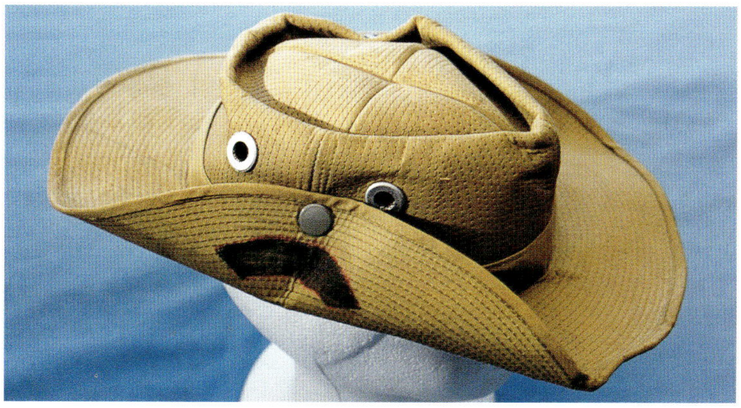

Close-up of tan cowboy hat worn by Ranger advisor.

Close-up of blue liner of cowboy hat. Note the number 6 inside.

Full view providing a good look at the Lieutenant's Ranger tab and lack of a shoulder sleeve insignia, or SSI. The other two American advisors on the team wore the BDQ Tiger patch as a shoulder sleeve insignia, instead of as a pocket patch or the MACV SSI.

Full figure view of advisor with captured Vietcong French M36 rifle.

Full figure of advisor examining a well used French M36 rifle. This also provides a good view of the tan cowboy hat worn by the advisor.

Second model jungle fatigues are worn which feature the covered buttons; all other aspects of the uniform stayed the same. On the jungle fatigue shirt is a full color U.S. ARMY strip, white name tape, full color white U.S. basic airborne wings, and a black and gold RANGER tab, showing he is a graduate of the demanding Ranger course at Fort Benning, Georgia. Pinned to his collar points are the gold crossed rifles (Infantry) and silver 1st Lieutenant bars (O-2). Pinned to his left pocket flap are two gold blossoms of the new ARVN rank, showing he is a Trung Uy, or Lieutenant. Usually, these would be worn on the front of the shirt, but in the Lieutenant's case, he has decided to secure them to his shirt flap, because his Red/Yellow scarf would obstruct the view of his rank. The two-tone Red/Yellow scarf was the Headquarters' Company's colors. Each of the rifle companies had different colored scarves, which made identification easy on the battlefield. The American advisors were almost always attached to the HQ element, and therefore would wear the HQ colors. The other Ranger Companies in the 42nd ARVN Rangers had maroon, forest green, and black.

Sgt. First Class David Williams with members of the 42nd ARVN Rangers, 3rd platoon. Note the members wear maroon scarves for identification. Each platoon had its own distinct color scarf. Williams wears the color of the Head Quarters element, which is yellow and red. Note the radioman on the far left and the man next to him with the captured VC trumpet. All men are wearing Olive Drab uniforms, except the man kneeling in front of SFC Williams, who wears tigerstripe camouflage fatigues. Note the man on SFC Williams' right is armed with a B.A.R. machine gun. (Dave Williams collection)

Three Ranger scrolls worn by American advisors only. 42nd BN, 39th BN, and 41st BN. All are hand sewn on ERDL pattern material. Note the three scrolls are on a piece of parachute material. (Bob Chatt collection)

Three Ranger scrolls of the Vietnamese Rangers. The top scroll is on a piece of the ARVN Ranger material, and the bottom one is done on black material.(Bob Chatt collection)

Left to right are Lieutenant Dale Shipley, Captain Robert Butler, and Sgt. First-Class David Williams posed for a photo before a mission, which may last up to three days. Note all three men wear a mixture of WWII era and M1956 gear. Bac Lieu, Vietnam, 1964-65. (David Williams collection.)

Captain Nguyen Tan Giai stands next to captured Viet Cong weapons, which include numerous M1 carbines and rifles, including German made MG. 42 and numerous other Chi-Com. weapons, 1964. Capt. Giai wears the Vietnamese made Olive drab fatigues; note the rank on his pocket flap. Weapons were captured on December 5, 1964, during Operations Dan Chi 92. (David Williams collection)

SFC David Williams at the advisor compound, Bac Lieu. He is wearing his duck hunter camouflage fatigues. Note the addition of a white name strip, color U.S. Army, pin-on CIB, pin-on Ranger qualification badge, and the addition of his rank, E-7, in full color on both sleeves. He also wears U.S. produced jump boots, a black trouser belt with brass roller buckle, and a Vietnamese made maroon Ranger beret. Standing to his left is Captain Giai in olive drab and two other unknown Vietnamese Rangers, 1964-65.

Full figure view of 42nd Ranger Advisor, Team 51, Bac Lieu, Vietnam, 1964.

Minimal web gear is worn by this advisor to include a M1956 pistol belt without the cotton H-harness, a twin magazine pouch for his WWII era M-1 carbine, an additional carbine pouch, which could accommodate four of the larger banana magazines, a single aluminum canteen with early cover, and a black leather holster for his .45 cal pistol. Additional gear, such as food and water, and occasionally poncho and liner, were carried in the advisor's box that was mounted to the HQ jeep, that usually went everywhere they did, so the advisors could travel light and without being burdened with rucksacks. 2nd model jungle boots are worn.

U.S. RANGER ADVISOR,
TEAM 51,
42ND ARVN RANGERS,
BAC LIEU, VIETNAM
1964-65

This advisor wears a mismatched set of duckhunter that has faded to pink and purple. CIA personnel operating in Southeast Asia in the mid-50s purchased this pattern from U.S. sporting goods outlets and wore them in Vietnam. As the war outgrew CIA control, the U.S. Government placed orders for uniforms with other Southeast Asian countries under the Mutual Defense Assistance Program (MDAP). Like most things, Vietnamese copies of these were soon available in several styles and sizes.

The camouflage is printed on heavyweight cotton cloth, which made it very uncomfortable in the intense heat and humidity of Vietnam. The shirt features two upper breast pockets, each secured with a single flat button, while the trousers feature two front (slash) pockets, two rear, and two side cargo pockets, which are rather large for Asian garments. The shirt is fully badged; something not often seen on camouflage uniforms, especially the duck hunter uniform. Worn above his right pocket is a U.S.-made silk black and gold U.S. ARMY strip, with a pin-on metal CIB. Just below the U.S. ARMY strip and slightly above the button of the pocket is the "Republic of Korea Presidential Unit Citation," awarded by the Korean Government in recognition of service during the Korean War. An individual assigned to a unit when they earned the citation may wear it as a permanent part of his uniform, but this decoration is not authorized for temporary wear by virtue of being assigned to a unit that has been awarded it.[104] On his right side he wears a color name strip, "Youngberg," along with a Vietnamese BDQ (Ranger Qualification Badge), honorarily awarded to the U.S. advisor. It features a wreath with crossed swords with a single star. It is believed that the BDQ patch and

Full figure shot provides a good look at the Team 51 advisor's duck hunter camouflage, complete with full color E-7 rank, nametape, and U.S. ARMY strip. All these items are usually not found on the Duck Hunters.

Full figure shot of Team 51 advisor to the 42nd ARVN Rangers at Bac Lieu.

Additional full figure shot of Ranger advisor. Note the duck hunter camouflage has faded to an off pink and purple color tones.

Upper torso shot provides a good look at the advisor's ARVN Ranger beret.

Close-up of advisor's right side with white nametape, BDQ qualification badge, and Presidential Unit Citation. Note these are items not usually found on camouflage or fatigue uniforms.

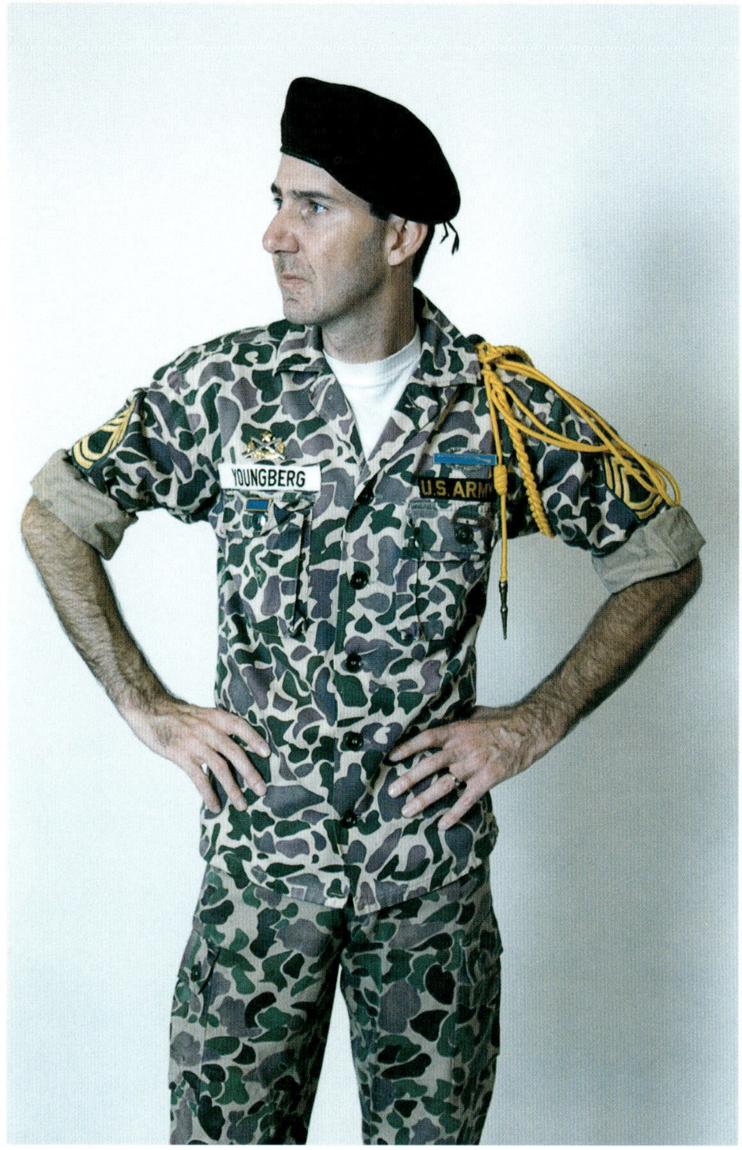

Upper torso shots provide a good look at the advisor's ceremonial uniform. This uniform was worn only at the compound, and was too hot and heavy to be worn on operations.

Close-up shows the ARVN beret with bullion badge and maker label.

Close-up of advisor's left pocket shows the Vietnamese yellow Fourragere cord, pin-on CIB, yellow on black U.S. Army, and Korean War Citation.

Three Vietnamese made Ranger scrolls. All are hand done on ERDL material. They are the 86th BN, 76th BN, and 85th BN.

qualification badge were adopted in 1961, being designed by LTC Millet, who was one of the first to receive the badge. Worn on the right pocket flap is the blue "Presidential Unit Citation," which is "awarded to units of the Armed forces of the U.S. and co-belligerent nations for extraordinary heroism in action against an armed enemy." The unit must display such gallantry, determination, and espirt de corps in accomplishing the mission under extremely difficult conditions as to set it apart and above other units participating in the same campaign. Only on rare occasions will a unit larger than a battalion qualify for award of this decoration.[105] Worn on his left sleeve is the yellow Vietnamese Fourragere;

Again polling French tradition for awards, the Fourragere resides by decree of 20 January 1953, the war to Vietnamese armed forces units and decided at the armed forces level two more times.

Sgt. First Class David Williams wearing Vietnamese produced "Duck Hunter" camouflage, which he obtained from the Vietnamese quartermasters. Note the black jump boots, the French Fourragere, ARVN Ranger badge, and the Presidential unit Citation worn on his right pocket flap. Headgear is the maroon ARVN Ranger beret. This uniform was only worn around the compound and not on operations, due to the fact that it was very heavy and hot. Team 51 Advisor compound, Bac Lieu, 1964-65. (Dave Williams collection)

Lieutenant Dale Shipley on an operation in the Delta with the 42nd Vietnamese Rangers. Uniform consists of first model jungle fatigues with a B.D.Q. shoulder sleeve insignia and Biet Dong Quan arc. Around his neck is worn the headquarters colors, red and yellow. Note the U.S. lightweight rucksack. Headgear is a locally purchased camouflage cowboy hat. (Dave Williams collection)

Close-up of maroon ARVN Ranger Beret.

Chalkboard showing weapons captured and Viet Cong killed during "Operations Dan Chi 92," December 5, 1964. The first line reads "Ton That Dich" and details 115 enemy killed during the operation. (Dave Williams collection)

These Fourragere followed the French pattern and were braided cords with the French knot at one and brass ferrule often called a pencil or hose nozzle at the other common and were worn over the left shoulder.

When an individual is assigned to a Vietnamese unit, such as the 42nd ARVN Rangers, he may wear it as a temporary part of the uniform only as long as he remains with that unit. Worn on both sleeves are full color U.S.-produced E-7 stripes, something rarely found on duck hunter uniforms. This was the NCO's ceremonial uniform, worn only in the confines of the base, and not worn on operations. These are sometimes referred to as "Christmas tree uniforms," "High Profile suits," or the "I love me uniform." Underneath is a white GI issue T-shirt.

Vietcong captured weapons during "Operation Dan Chi 92." Note the numerous German made MG 42s on the anti-aircraft tripods, along with the numerous sub-machineguns. (David Williams collection).

Sign outside the advisory team 58 compound in Chuong Thien. Note the MAAG Vietnam symbol is still in use, although the team was now under MACV control—nobody wanted to take the time to change the sign over to MACV. Team 51 spent some time here during operations and met the actor Raymond Burr. (David Williams collection)

Captain Robert Butler (right) with Captain Giai (left). Capt. Butler wears first model jungle fatigues with exposed buttons; note the full-color U.S. Army, white name strip, and Vietnamese Ranger badge on sleeve with full-color Ranger tab. Pinned to his left pocket flap is his Vietnamese rank, Dai Uy, along with a red/yellow scarf and maroon Vietnamese Ranger beret. Capt. Giai wears duck hunter fatigues underneath his M1951 field jacket. Note the large star-shaped Ranger patch on the M1951 field jacket, 1964. (David Williams collection.)

All three members of team 51 return from an operation in the delta, 1964. From left to right are LT. Dale Shipley, Sgt. First Class David Williams, and Captain Robert Butler. All three men wear exposed buttons fatigues and camouflage cowboy hats. Around their necks they wear the red and yellow scarves of the Headquarters Company. All three are lightly equipped with a mixture of Korean War and WWII gear. Note that Capt. Butler does not wear suspenders with his pistol belt. All are wet from the waist down, showing prolonged operations in the delta. Circa 1964. (David Williams collection.)

Headgear consists of the maroon ARVN Ranger beret, which was one of the most distinguishing items of the ARVN Rangers. The colorful Biet Dong Quan berets were usually garnet in color, but varied sometimes, being orange, rose, or maroon. U.S. Army Advisors began wearing them officially in 1965, but many advisors had been wearing them as early as 1962. This one features the gold bullion beret badge. These berets were usually awarded to the advisor during a lavish ceremony, which made them part of the Ranger family. Black leather jump boots complete the advisor's uniform.

Sgt. First Class David Williams (right) wears standard issue of first model OG 107 cotton sateen fatigues. Note the lack of the MACV patch on his left shoulder, and around his neck is the yellow and red headquarters scarf. On his head he wears the maroon BDQ beret. Note the small Korean War Citation on his pocket flap. The man standing next to him is a Cambodian serving in the Vietnamese Rangers. He wears a strange pattern of leopard camouflage with a matching rain jacket and color scarf. (David Williams collection.)

Captain Giai and Sgt. First Class David Williams stand in front of a headquarters jeep, which would accompany the team to the field and was used to carry food and other essentials. Capt. Giai wears duck hunter fatigues, a camouflage scarf, and field jacket. Note the ARVN Ranger tiger patch on the left pocket of the field jacket. SFC David Williams wears 1st model OG 107 cotton sateen fatigues with the first model jungle boots, full-color rank, and U.S. Army and name strips. Note the Korean War Citation worn on the left pocket flap, along with the B.D.Q. patch worn on the left sleeve. Headgear is the Vietnamese Ranger beret. Around his neck is the two-tone red and yellow HQ company scarf, 1964. (Dave Williams collection)

U.S. RANGER ADVISOR
TEAM 51,
42nd ARVN RANGERS
BAC LIEU, VIETNAM
1964-65

An Xuyen Province, Vietnam-SFC David M. Williams, a member of 51st Advisory Team assigned to the 42nd ARVN Ranger Battalion, has just returned from a mission where the results were an afternoon raid against the Communist Viet Cong. Arriving in the second of four Army heliborne "Eagle Flights," SFC Williams' unit began forcing the VC westward. The Vietnamese Rangers crossed a nearby canal and the VC withdrew. At the same time, another "Eagle Flight" landed behind the retreating communists, catching them in a vise. SFC Williams recalled, "that without the helicopter support we had during the operation we would have suffered many more casualties than we did." Over 115 Communists were killed, while Government troops suffered 29 dead.

Full front view of SFC Armstrong, Team 51, Bac Lieu compound. Advisor to the 42nd ARVN Rangers, 1965.

Close-up of advisor provides a good look at the advisor's bright yellow and red scarf, second award Combat Infantryman's Badge, Presidential Unit Citation, and a very well made BDQ qualification badge.

Full figure view, with emphasis on the advisor's left side, provides a good shot of his B.D.Q. shoulder sleeve insignia.

Full view of Team 51 advisor to the 42nd Rangers provides a good view of the advisor's red and yellow scarf.

Upper torso shot of Team 51 advisor provides a good view of the advisor's black belt with the gold roller buckle. This also provides a good view of his Bronze Star and strange tiger stripe pattern cowboy hat with an extremely large brim.

Nov. 14, 1964, Viet IV Corps Awards Medals to 15 Americans: 15 Americans were awarded Vietnamese medals at ceremonies held recently in the 21st Vietnamese Division. Presentation was by Gen. Nguyen Khanh, ARVN Commander in Chief, and the commanding generals of the IV Corps and the 21st Division. Receiving the Vietnamese Army Medal of Valor (Army Level) were Col. Jim Kersey, Maj. Levinson, and Captains Ramsey, Gordon, Nye, Hamner, and Kemdrick. The medal of Valor (Corps level) was presented to Lt. Col. Lee, and Captains Shoemaker, Henderson, Scharge, and Robert Butler. The second award in less than 60 days of this medal was also received by 1st Lieutenant Dale Shipley. The Vietnamese Army Medals of Honor were also presented to Capt. Charles W. Brown and SFC David Williams.

This U.S. Advisor displays the diversity of uniforms worn by members of the 42nd in 1965. He wears the 1st model OG 107 cotton sateen fatigues, which were first produced in 1952, but soon would be improved by altering the button spacing and re-aligning the pockets to produce a higher neck opening. In addi-

SFC Dave Williams being awarded the Bronze Star during award ceremony at Bac Lieu in 1965. He wears first model OG 107 cotton sateen fatigues with a second award CIB. The man shaking his hand is Col. Barton, Senior Advisor. (David Williams collection.)

Sgt. First Class David Williams enjoys a coconut during a lull in the operation in the province of An Xuyen Province, South Vietnam. Note the red and yellow scarf worn around his neck, 1964. (David Williams collection)

Three Vietnamese Ranger scrolls from the 2nd BN, 11th BN, and 21st BN.

tion, the shirt would feature "clipped cornered pockets" in conformity with other Army shirts. The first model shirt is seen here tucked into the OG 107-cotton trouser and accompanied with the black web belt and gold roller buckle. Worn above his left pocket is a U.S. ARMY strip in the branch colors, gold and black, and a pin-back combat infantryman's badge, second award, which features a star. Many advisors followed their contemporaries and wore ribbon bars and Qualification Badges in "secure areas," and during Award ceremonies, changes of command or when deemed necessary by higher ups. These could easily be removed before going to the field and placed in one's pocket, then put back on once back in the security of the compound. Above his name strip on the right side is a well made BDQ Qualification Badge. Below, on his pocket flap is the Presidential Unit Citation. Sometimes a Battalion patch, usually round and printed on cotton, would be worn on the adjacent pocket along with the BDQ patch. This advisor has chosen to wear his silk woven Ranger patch as an SSI Shoulder Sleeve Insignia, instead of the MACV patch. Note this Ranger advisor has just been awarded the Bronze Star Medal for heroism. Worn around his neck is the Headquarters Company Red/

Sgt. First Class David Williams (left) and Captain Robert W. Butler (right) display captured Viet Cong flag after a successful operation in the Delta. Both men wear first model jungle fatigues with exposed buttons. (Dave Williams collection.)

Sign outside of Bac Lieu. This sign was a bit of a joke; note the field elevation during the wet season and the dry season. (David Williams collection.)

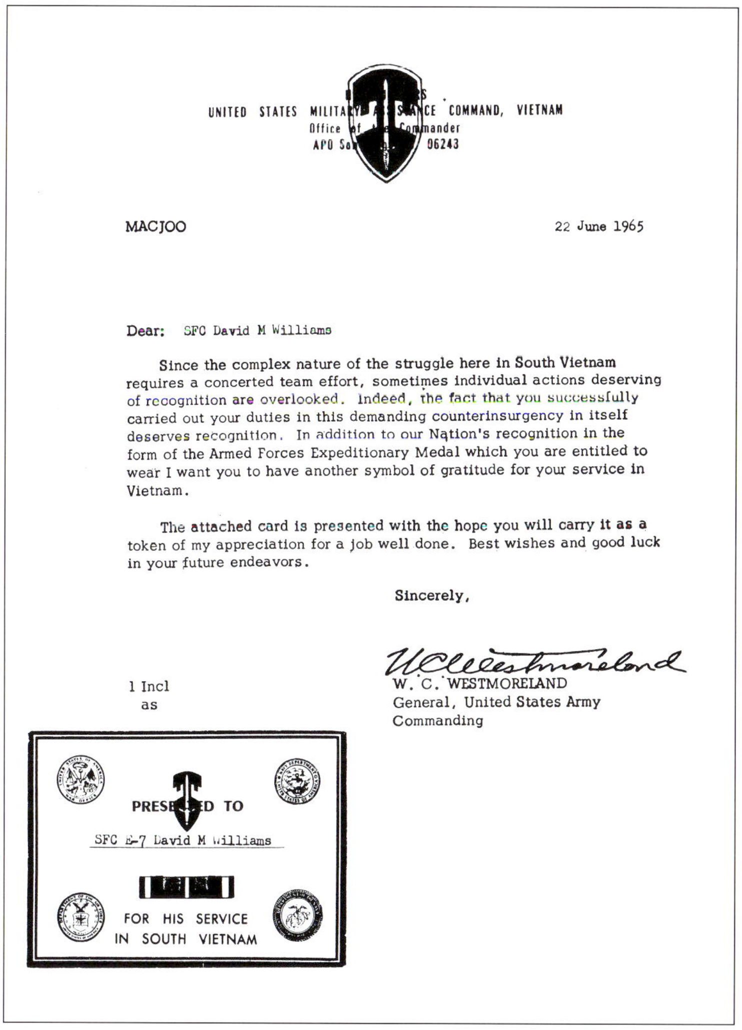

UNITED STATES MILITA… COMMAND, VIETNAM
Office … Commander
APO … 96243

MACJOO

22 June 1965

Dear: SFC David M Williams

Since the complex nature of the struggle here in South Vietnam requires a concerted team effort, sometimes individual actions deserving of recognition are overlooked. Indeed, the fact that you successfully carried out your duties in this demanding counterinsurgency in itself deserves recognition. In addition to our Nation's recognition in the form of the Armed Forces Expeditionary Medal which you are entitled to wear I want you to have another symbol of gratitude for your service in Vietnam.

The attached card is presented with the hope you will carry it as a token of my appreciation for a job well done. Best wishes and good luck in your future endeavors.

Sincerely,

W. C. WESTMORELAND
General, United States Army
Commanding

1 Incl
as

PRESE[…]D TO
SFC E-7 David M Williams
FOR HIS SERVICE
IN SOUTH VIETNAM

Letter to David Williams from General William Westmoreland, dated 22 June 1965. (David Williams collection.)

Two advisors assigned to the Duc My Ranger school put a .30 cal. Machine gun into action, 1965.

Yellow identification scarf. The three American advisors to the 42nd (SFC Williams, Lt. Shipley, and Capt. Butler) always wore the Red/Yellow scarf, even if operating with another company. The Rangers favored these scarves; their ranks were full of Cambodians, and these brightly colored scarves were symbols of religion, medicine, and good luck—much like those worn by the American Indians.

On his head is a slightly worn Vietnamese tigerstripe cowboy hat with a large brim. Once again, black leather jump boots complete the advisor's uniform.

U.S. RANGER ADVISORS, DUC MY RANGER TRAINING CENTER, 1965

In 1961, Ranger training was conducted at two sites during this period, Trung Lap in III CTZ (War Zone D, near Cu Chi) and Tot Son in IV CTZ. Shortly thereafter a Ranger Training Center was established at Duc My. The American Advisors with select Vietnamese Cadre would take these newly-designated Ranger units through the ranger counter-guerrilla course; the last two weeks would be out in the field operating against a live enemy. The Special Battalions were the 10th Battalion, which was for I Corps, located at Da Nang; the 20th Battalion for II Corps, located at

Author, Gentleman, Professional Soldier, and Scholar CSM Mike Martin with the 44th ARVN Rangers, Soc Trang, 1965.

With the machine gun in place, the advisors explain the proper way to load the machine gun.

Note the two advisors wear the 1st model exposed button jungle fatigues and the ARVN Ranger patch on their pockets.

Pleiku; and the 30th Battalion for the III Corps, located in Saigon. The first Ranger battalions were formed and trained at Trung Lap in the summer of 1963, the 32nd Ranger Battalion being the first battalion to go through the training and actually operate as a Ranger unit.[107]

The Duc My Ranger camp operated on the same lines as the U.S. Ranger School. Both it and the Trung Lap school conducted a type of refresher course for all of Vietnam's Ranger battalions. This training was from two to three months in duration, with the battalions receiving basic and unit level training with emphasis being actual combat operations. Units also received new equipment and uniforms at the camps. It was a kind of break and rejuvenation period for the battle-scarred Rangers. The objective of Ranger training was to develop in each soldier pride, confidence, self determination, and the ability to fight and succeed regardless of the enemy, odds, weather, or terrain.[108]

This reconstruction of two U.S. advisors is representative of any of the men who taught and lived with the Vietnamese Rangers at Duc My Training Center in 1965. Both Ranger Advisors wear the first model jungle fatigues with exposed buttons. The jungle fatigues were first introduced in 1963 and combined comfort, practicality, and a smart appearance. The Tropical Combat uniform, or "Jungle Fatigues," was to be the primary field uniform in Vietnam after 1963. This uniform was a rare example of

Ranger advisor shows the Vietnamese at the Duc My Training Center the use of the M-1 Garrand rifle grenade. Note the Vietnamese made black cowboy holster and the BDQ patch on the pocket. Note also the small ARVN Ranger beret.

The two Ranger advisors at the Duc My Ranger Training center. Note they both wear the ARVN Ranger berets.

Two ARVN Rangers both wearing the ARVN Ranger pattern fatigues with the BDQ on their left sleeves, 1965-66.

Army clothing that combined functional practicality with comfort and an attractive design. The quick-drying coat and trousers were made of a tightly woven lightweight twill cotton fabric. The loose fitting garments promoted body and heat ventilation and moisture dissipation. The coat was designed to be worn outside the trousers, and sleeves were allowed to be rolled up. In fact, it became mandatory to roll up the sleeves above the elbows neatly in military installations and rolled down while in the fields. The coat features two slanted chest pockets, which allowed soldiers ease in removal of items, and two lower bellow pockets. The shirt also incorporated shoulder loops, side button take-up tabs, and an "inner gas flap." All buttons on the garment are "exposed."[109] The trousers also featured the "exposed" buttons to match the shirt. They featured two hip pockets, two front pockets, and two large bellow cargo leg pockets. Inside the cargo pockets were two integral tie-down straps, much like the M1951 field trouser offered. Inside the left cargo pocket was a small one-button pocket that was designed to hold a small survival kit that was never satisfactorily developed or issued.

The Advisor wears full color U.S. insignia, along with a pair of South Vietnamese jump wings and a BDQ patch on the pocket. Headgear worn by both Advisors is the maroon Ranger beret first authorized for wear by U.S. personnel in 1965, although some advisors had been wearing them as early as 1962. This Advisor wears the "bevo" woven variation of the arrow and wreath. Around his waist is a Vietnamese "cowboy rig" with a quick-draw holster for his M1911 .45 pistol. He holds the M1919 Browning Machinegun with the T.O. & E. attached.

(Right) The second advisor wears essentially the same uniform, minus the Vietnamese wings. Across his chest is the russet brown M3 WWII holster for his M1911 .45 pistol, worn more for effect and the mark of authority. His BDQ beret is a cherry red color with an early beret badge, which is a crude tin metal Arrow, Wreath, and Wing beret device. In front of the two Advisors is the M2 tripod, a can of linked .30 cal. ammo, and a TM for the M1919. The second Advisor wears full color U.S. insignia, along with a pair of South Vietnamese jump wings and a BDQ patch on the pocket. Headgear worn by both Advisors is the maroon Ranger beret first authorized for wear by U.S. personnel in 1965, although some advisors were wearing them as early as 1960. Both advisors wear the second model jungle boots, which lack the ankle support and the slip-in spike protector.

Two ARVN Ranger shows the diversity of uniforms worn. One wears tigers, while the other appears to be an officer wearing the early pattern ARVN Ranger pattern with its washed out tones, along with a para silk scarf. Note the Vietnamese in tigerstripe wears a pair of Bata boots.

Full figure shot of the American advisor at the Duc My Ranger Training Center, 1966.

U.S. RANGER ADVISOR, DUC MY RANGER TRAINING CENTER, 1965

This advisor at the Duc My Training center is shown here instructing the ARVN Rangers in the use of the M1 rifle, which is fitted with an M7A3 grenade launcher, along with the M15 grenade sight. He has mounted an M19A1 anti-tank rifle grenade to the launcher for use against suspected VC fortified bunkers. Standard MKIIA1 pineapple hand grenades could be launched this way, also, with the addition of an M1A2 grenade projection adapter.[110]

The Advisor wears full-color U.S. insignia, along with a pair of South Vietnamese jump wings and a BDQ patch on the pocket of his first model exposed button jungle fatigues. Headgear worn by the Advisor is a rather small, maroon ARVN Ranger beret. Second model jungle boots are worn here.

Right: *Upper torso shot provides a good view of the ARVN Ranger advisor's beret flash. The shot also provides a good view of the small pocket on his right sleeve, along with his pin-on rank, Captain, on his collar. Many U.S. advisors at the Duc My Ranger Training Center wore the Vietnamese Ranger scroll on the right sleeve that read "Duc My Training Center," although the schools' scroll is not worn by this advisor.*

Full figure shot showing the ARVN Ranger advisor with emphasis on his right side. Note the black leather holster worn as a symbol of authority more than as personal protection.

Upper torso shot showing the Ranger advisor's U.S. Army tape, "Master Blaster" wings, and Combat Infantryman's Badge. Note the pocket flap is patterned much like the U.S. made second model OD 107 cotton sateens. On the collar are pin-on color Captain bars. Note the pen pocket on his left sleeve just below the Ranger BDQ tiger Shoulder Sleeve Insignia. This provides a good view of the ARVN Ranger beret with the bullion badge.

U.S. RANGER ADVISOR,
Ranger Training Center, Duc My
1965-66

The Ranger Training Center at Duc My thoroughly trained Ranger soldiers to accomplish the difficult and varied missions assigned to Ranger units. Known as the "steel refinery center" of the ARVN, the center conducted training in both jungle and mountain warfare. Graduates of the school were awarded the ranger badge. The center also trained the Rangers and other services in the techniques of Long-Range Reconnaissance Patrolling, including Commando and Reconnaissance (RECONDO) training. Graduates of this course were awarded the winged binocular badge.[111] Army advisory duty was one of the most dangerous assignments in Vietnam. Service with Vietnamese units began to invite selective VC reprisal during 1963, and advisor casualties increased to alarming levels in 1964.[112] Due to an alarming increase in U.S. Advisor casualties in 1964, it became obvious that the Viet Cong were deliberately targeting Americans. On 2 July 1965, MACV directed U.S. Advisors to begin wearing South Vietnamese Military uniforms as a means of "personal camouflage." Wearing of the Vietnamese uniform was recommended during operations outside the safety of the compound. In addition, it was advised that they wear the uniform, in whole or in part, when living in isolated areas with the Ranger units, or when the U.S. Advisors would be clearly distinguishable from Vietnamese personnel.[113] This jaunty Ranger Advisor (1965) wears the complete uniform of his host unit, including the Biet Dong Quan beret. The camouflage uniform worn here came to be known as the ARVN Ranger pattern, although other elite units would utilize it also. A direct copy of the U.S. E.R.D.L. pattern, it was first introduced in late 1962 by the U.S. Combat Development Test Center, Vietnam. Printed on both lightweight poplin and mid weight cotton, as seen here, and later heavy weight. The uniform featured two breast pockets, reinforcements on the elbows, and two pen pockets, one on each of the sleeves, typical of Vietnamese garments. This pattern was produced in several weights and versions throughout the war; later, it would take on a darker color scheme, but nevertheless, always recognizable as a version of the original. The shirt worn here is an example of the early pattern and is fully badged. On his left sleeve is the Vietnamese Ranger SSI, which is silk woven. Worn above

Duc My RTC scroll (center) would be worn by the Vietnamese Ranger advisors, along with a scroll from the 30th and 40th Recon.

Three Ranger scrolls from the I Corps, High Command, and the 77th Recon, as evident by TM. All are on ERDL material and are hand sewn.

his Vietnamese-made subdued U.S. ARMY strip are subdued Master jump wings and a subdued CIB, very unusual for this early in the conflict. Both his rank (Captain) and crossed rifles are U.S. made. Pants are printed with the same pattern and feature two rear pockets and two large cargo pockets on his hips similar to WWII style, but lack the front slash pockets.

The Biet Dong Quan beret, worn here, is of a deep garnet color, but varied sometimes, to orange, rose, or maroon. On the beret is the winged arrow with a wreath done in gold bullion. The Vietnamese, along with their U.S. Advisors, wore the beret "French" style, pulled over the left eye, with the badge over the right eye, or as in this case, his right ear. Many U.S. advisors would wear their U.S. rank or Vietnamese equivalent affixed next to the beret badge. His black leather holster and twin cell magazine pouches are the only things on his M1956 pistol belt with the new Davis quick fastener. In his holster sits his 1911 .45 pistol. The tropical combat boots, or "jungle boots," were first introduced into Vietnam in 1962. This early pattern, made by the Bata Shoe Co., lacked the spike-protective insole and the reinforced ankles.

Close-up of ARVN Ranger maroon beret.

U.S. ARMY RANGER ADVISOR,
ARVN RANGER TRAINING CENTER,
Trung Lap
1966

This Ranger advisor stationed at the Trung Lap Ranger School wears a mismatched set of the First model "jungle fatigues." The Tropical combat coat and trousers, or "Jungle fatigue" as they were known, were made of lightweight 5.5 ounce cotton poplin fabric in Olive Drab green shade 107. The coat was a loose fitting "Bush jacket" style that was fitted with side take-up tabs and shoulder loops[114] The take-up tabs gave the uniform top a slimming appearance. The two upper bellow pockets were in slanted position, and the lower pockets were straight. The 1963-64 pattern top and bottoms had all exposed buttons, something that would be done away with by the 26th of August 1964. Note this advisor wears the sleeves down, which means it must be the cooler time of the year. On the coat are an early, silk woven, subdued U.S. ARMY and a locally-produced name strip. Worn above his U.S. ARMY strip are a set of Vietnamese-made "Master Blaster" airborne wings and a full-color hand sewn CIB. Worn on the collar points are a full-color yellow crossed rifles, signifying his Branch of Service (BOS) is Infantry. Worn on the right collar are the railroad tracks, signifying he is a Captain (0-3). Worn on the right pocket is a crudely-made ARVN Ranger BDQ pocket hanger,

Full figure shot of U.S. advisor assigned to the ARVN Ranger School at Ranger Training Center located at Duc My, 1966.

Full figure view, showing the advisor consulting a map of the training area before a training mission.

Full figure view with emphasis on the American advisor's right side. Note the civilian Thermos® with the Captain's "lifer's juice," i.e. coffee.

which is nothing more than a colored piece of paper with the panther head encased in plastic. Many different variations of these existed, some having actual silk woven patches, while others had cloth printed variations. Worn on the left sleeve is a well-made, machine-embroidered MACV SSI. His personal side arm, a Colt .45 1911A1 pistol, is secured in a WWII era russet brown shoulder holster, worn more for a sign of authority than personal protection. Headgear consists of the ever-popular tan "cowboy" boonie hat with a red and white arc that reads "Sorry about that Viet Nam." These were very popular and could be purchased at any of the local tailor shops that catered to the Army, and could make a cowboy hat in any pattern with anything written on an arc for a few cents to a dollar. second model DMS jungle boots are worn here. Note the advisor carries a map of the training area and a private purchase Thermos for his "lifer's juice," i.e. coffee.

Close-up of white on red arc.

Upper torso shot provides a good view of the ARVN Ranger B. D. Q. pocket hanger. This is also a shot of his M1911 A1 secured in the WWII era russet brown shoulder holster.

Close-up showing the advisor's U.S. Army tape, Vietnamese made full color master wings, and a hand sewn, full color Vietnamese made Combat Infantryman's Badge. This also provides a good view of the red and white arc on his cowboy hat, which reads "sorry about that, Vietnam." Note the Vietnamese made MACV SSI, which is subdued.

Three Vietnamese Ranger scrolls of the 1st GP, 35th BN, and 34th BN. Note the top one is slightly maroon and the other two are red. All three are on ERDL material.

III Corps third Ranger group, 31st Ranger battalion is different than the others, and features a red No. 3 with a white box of maroon tab with a white 31. Note the patches are subdued and silk woven. Silk woven, subdued Ranger patches are very rare. To the right of it is the 4th Ranger Group with yellow tab, bordered in black red No. 4 and white 42, attached to the patch. Note the size difference between the two.

U.S. RANGER ADVISORS,
44th ARVN RANGERS,
SOC TRANG, VIETNAM
1966.

These two Advisors both wear the second model poplin shirt with exposed buttons on the shoulder loops, and take-up tabs only. The buttons on the pockets have been concealed beneath a flap to prevent snagging. The jacket incorporated the same design and cotton material as the first model. The advisor on the right wears all color insignia, along with a full-color silk woven BDQ patch on his sleeve. Although obscured by his cotton H-harness, this officer wears pin-on BOS and Lieutenant rank on his collar points, which was not uncommon at this time. Note also the white plastic C-rat spoon in his left pocket. Trousers worn by both advisors are the second model jungle fatigues, which are identical to the first model fatigues, but by 1965 the buttons on the pockets were concealed beneath a flap to prevent snagging. The large cargo pockets retained the leg tie-downs, and the button take-up tabs along the waist band.[115] LBE consists of a M1944 pistol belt connected to a pair of cotton H suspenders, two universal pouches with magazines for his M-1 carbine, and two 1qt canteen covers with two M1910 canteens.

On his back is the ARVN "Ranger Pack," or indigenous rucksack (copied after an early two-pocket NVA pack), developed by the U.S. for the Vietnamese Rangers. It consisted of a large green cotton duck pack with two large outside pockets affixed to a steel X-frame. His M1 steel pot is fitted with the WWII helmet net. Many U.S. Advisors wore the same painted helmets as did the ARVN Rangers, but by the early part of 1966, the Ranger high command ordered the Rangers to stop putting the tiger on their helmets. Their reasoning was that the V.C. were fleeing at the sight of the Black Tigers. The order was never fully carried out, and the Rangers continued to wear the tiger proudly on their helmets, while others mildly attempted to cover the tigers with a WWII helmet net.

The Advisor on the left wears virtually the same uniform, but with a mixture of both subdued and color insignia. Both his nametape and U.S. Army strip are subdued, while his airborne wings MACV patch and Ranger tab are in color. Subdued insignia was authorized for all field and work uniforms in late 1966, but the changeover period would take a while, and it was not uncommon to see a mixture of full-color and subdued insignia together. Web gear worn by this Advisor consists of a cotton H-harness connected to an M1956 pistol belt with two universal pouches, an M4 bayonet in the M8 scabbard for his carbine, along with a single M1956 canteen cover and plastic 1qt canteen.

On his head he wears the 1951 style "patrol cap." Note, his M-1 helmet rests on his knee with a slightly larger pattern netting than his Lieutenant.

Worn here by both men are the new second model Tropical Combat boots, more commonly known as the "Jungle Boots," that were officially adopted in Jan. 1965. They featured a cotton/nylon upper which were cool and quick drying, a cleated sole, and two-screened eyeholes for drainage. The design was so successful, that other than an added strip to reinforce the ankles, a slip-in "spike protector," and a new anti-mud sole, the boot would remain unchanged and in service with the armed forces for 25 more years. Both men are armed the M-1 Carbine. On the ground in front of them is a map of the area, a GI Angle-head flashlight, an M1911 .45 pistol, and a lensatic compass.

Two U.S. Ranger advisors assigned to the 37th BDQ consult a map during operations in 1966. Both advisors wear the 2nd model jungle fatigues and carry M-1 carbines.

Close-up of well worn ARVN Ranger beret with hand sewn beret badge.

Portrait of Ranger Advisor. Note he wears the early washed out ARVN Ranger pattern with a full color silk woven BDQ Tiger head, a pin-on CIB, subdued U.S. Army, a U.S. made printed name tape (Williams), and Vietnamese made "Master Blaster" (Senior) wings. Headgear is the Maroon ARVN Ranger beret with the BDQ beret badge, his rank, and a small "beer can" ARVN Ranger D.I.

Three Ranger scrolls from the 67th BN, 66th BN, and the 75th BN. All three are hand sewn.

Young Vietnamese Ranger wearing a tiger stripe uniform with a full color BDQ, white T-shirt, and handsewn beret badge on his maroon beret.

RANGER ADVISOR,
42nd BDQ,
BAC LIEU, VIETNAM
1966

This advisor wears the third model, olive green shade 107 utility shirt, first adopted on 6 Nov 1964 and made of cotton sateen. It was designed with patch pockets and pointed flaps, newly-designed buttoned cuffs, and the new style dull plastic concave buttons. Both his name strip and U.S. Army tape are the new subdued version. He also wears a ribbon bar with the U.S. Vietnam Service ribbon, Cross of Gallantry, and the Republic of Vietnam Campaign Medal (white and green). Above that is a U.S. made C.I.B. Contrary to popular belief, U.S. Advisors to the ARVN Rangers were not all Airborne or Ranger qualified. Above his name is a sandcast BDQ qualification badge, which was awarded to the American advisors after being with the unit for a certain amount of time, and seeing some combat with the unit. A large gathering usually accompanied the award ceremony. Below his nametape is the Presidential Unit Citation. Very few U.S. units were ever awarded this, and even fewer Vietnamese units. A full-color, bevo silk woven BDQ patch is worn on the right pocket. Full color E-7 stripes (Sgt. First Class) are worn on both sleeves.

Full view of Ranger advisor at Bac Lieu, 42nd ARVN Rangers, 1966.

Three quarter front left view of 42nd Ranger advisor.

Front view of 42nd Ranger advisor at Bac Lieu, 1966.

Upper torso shot of ARVN Ranger advisor. Note the Tin beret badge is not flush against the maroon beret.

Upper torso shot showing beret badge lying flush against the beret.

Three quarter front right view of BDQ advisor. Circa 1966.

Close-up showing American advisor's right side. Note the Presidential Unit Citation pinned to the advisor's pocket flap, and the ribbon bar, an item usually not found on a fatigue uniform. This is also a good view of his private purchase airborne brass roller buckle, as well as a good look at the sand cast BDQ badge.

Close-up of American advisor and his maroon ARVN beret. Note the early Tin badge is still in use in 1966.

Close-up of the advisor's left side. This provides a good view of the ribbon bar with the Vietnamese cross of gallantry, which was awarded for heroism and came in four levels of awards. The ribbon bar would be taken off before operations and placed in the advisor's pocket. This is a good shot of the advisor's Second Award of the Silver Star. Note the "Skillcraft" ball point pen in his pocket.

Another upper torso close-up showing the advisor's left side, which provides a good view of his MACV SSI and the E-7 chevrons.

Full view of ARVN Ranger advisor in 1966.

Atop his head is a Vietnamese Ranger beret with an early tin arrow and wreath badge. Second model jungle boots are worn here.

U.S. RANGER ADVISOR,
52nd Ranger Battalion,
1966

This advisor to the Vietnamese Rangers wears a set of tiger stripe known by collectors as the gold pattern. The pattern is printed on heavyweight cotton material, which takes on a gold hue after being worn and washed a few times. It features two upper breast pockets, each secured with a single button, and buttoned cuffs. This pattern usually features a small cigarette or first aid dressing pocket on the left sleeve. The matching trousers feature two rear pockets, two cargo pockets, and a small cigarette pocket on the left side. These trousers lacked the front slash pockets commonly found on American garments. Worn on his tiger stripe shirt are a set of full-color E-6 stripes that are U.S. made, along with a silk woven Vietnamese BDQ SSI.

LBE consists of an M1956 pistol belt with the three M1956 universal pouches, and a single plastic Canteen with the M1956 Canteen cover. Connected to his pistol belt are the cotton H harness suspenders with a single D-ring and first aid/compass pouch. On his back he carries the PRC-25 radio on the harness electrical

U.S. advisor utilizes the radio to call in support for the ARVN Rangers.

Rear view of advisor provides a good look at his PRC 25 radio on its pack board.

Full view of ARVN Ranger advisor.

Close-up shows the Rangers utilizing the AN/PRC 25 radio.

Full left side of Ranger kneeling.

Another shot of the ARVN Ranger advisor utilizing his radio.

5th Vietnamese Ranger Recon scroll (Top), along with the 6th Recon and the 7th Recon (bottom). The top two are on black material, while the bottom is on ERDL material.

equipment (ST-138/PRC-25), otherwise known as the carrying assembly, which is made of cotton duck. It is used to secure the PRC 25 so that it can be carried on the operator's back. The radio is usually issued with the following: bag cotton duck, or accessory bag, which is sectioned into several pockets, which are used to store the two antennas and handset H-138; supports; antenna, antenna AT-271A/PRC (long whip); and antenna AT-892/PRC-25 (short whip). The battery is the standard carbon dry cell battery; BA-4386/PRC25/U is a magnesium cell battery that would operate the radio approximately twice as long as the BA-386/PRC 25.

Close-up shows the Ranger's tigerstripe uniform, BDQ shoulder sleeve insignia, and PRC-25 radio.

Personalized Zippo lighter with BDQ tiger head.

2nd Vietnamese Ranger RCN, along with the 3RD RCN and the 4th RCN.

Upper torso shot of 42nd ARVN Ranger advisor, Bac Lieu, 1967. He wears the Army tan Class B uniform with the ARVN Ranger beret.

Headgear is the M1 steel pot with an early Mitchell pattern camouflage cover. The third model DMS jungle boots are worn, which feature the slip-in spike protector and ankle reinforcements. In his hand he carries his M1 carbine.

U.S. RANGER ADVISOR,
42ND ARVN RANGERS,
BAC LIEU, VIETNAM
1966-67

Worn by this Advisor is the short sleeve khaki uniform shirt that was "produced according to specifications of 19 December 1955." The shirt was introduced as a component of the abbreviated khaki cotton uniform. On 01 July 1956 it became a mandatory item of khaki attire commencing 01 July 1958. The shirt was made of 8.2-ounce cotton twill in khaki shade 1. It was designed with an open stock collar without a neck button. The shirt has shoulder loops, two box-pleat pockets with buttoned pointed flaps, and a square bottom with side vents at the hem. Later it was modified to simplify production and to conform to Army style conventional shirt patterns. The box pleat had pointed flaps, or was eliminated

III Corps. Six Ranger group 35th and 34th Ranger battalions.

Three Vietnamese Ranger scrolls, featuring the 5th GP (Top), 6th Group (GP), and the 7th Group (GP). All three scrolls are hand sewn on ERDL material.

From left to right; "4th Ranger group, 44th Ranger battalion machine sewn on the ARVN Ranger camouflage material with a yellow tab, and are 4 and two white 44s. 4th Ranger Group tab on a yellow tab bordered black, are a red No. 4 within a red border white square. The battalion numbers 42, 43, and 44 are white, always attached to the insignia, and machine sewn on ARVN Ranger camouflage material." (S. Smyth)

from the pocket design, as seen here.[116] The conventional khaki uniform trousers, worn here, were also made of 8.2 ounce cotton twill in khaki shade 1. The design was standard, with slip-type hip pockets, a watch pocket that was later eliminated, and buttoned or slide-fastened fly closure, i.e. zipper. The first pattern trousers had a self waistband with the narrow belt loops and buttoned fly.[117]

After 1948, the waist belt was the olive drab shade 3 web belt with GCM buckle. Khaki or Olive green web belts were optional. On 29 August 1957 officers were directed to wear a black web or optional woven elastic belt. On 13 June 1958 the belt became available for enlisted wear, except in formation. On 01 Oct. 1958 the belt became part of the initial clothing issue, and on 15 April 1959 the black belt officially replaced the olive drab belt.[118] On the left side of his tan service uniform the advisor wears basic airborne wings on a blue Airborne oval. Worn beside that is his expert badge showing he is an expert in the rifle, pistol, and machine gun. Decorations worn on his left side include, from top to bottom and most important to least important: Army Commendation; Good Conduct; American Defense; Armed Forces Expeditionary Medal; Vietnam Service Medal; Vietnam Gallantry Cross; and the Republic of Vietnam Campaign Medal. The Republic of Vietnam Campaign Medal, which is green and white, was awarded by the government of the Republic of Vietnam to members of the United States Armed Forces and authorized by department of Defense directives 348.17, 20 June 1966. To be eligible, an individual must have served in South Vietnam for six months during wartime, or meeting criteria established for the armed forces expedition area medal. Vietnamese wore the Vietnam service medal for those directly supporting the war in the mother countries. An individual could also qualify if he was wounded by hostile forces, captured and later rescued or released, or killed in action (Republican Vietnam campaign medal was not issued, all are service ribbon with the 1960 device). Next to that is unknown. The level awarded, indicated by a different apparatus attached to the suspension of service ribbon, was predicated upon the degree of heroism involved. The high grade, with Palm frond, was presented at the national level. The Gold-star indicated Corps level, a Silver Star indicated division level, and a broad star represented the regimental level (Therefore, the Gallantry Cross of the silver star was approved by Norman division commander or equivalent and presented at a division review). The appurtenances of subsequent awards of the gallantry cross were attached to the suspension or service ribbon, and medals or service ribbons added when they were too numerous to fit on one ribbon. On the third row, from left to right are the Vietnam Service Medal, which is yellow with three red stripes and green corners. This was established by executive order 11231 and is authorized for all members of the armed forces of the United States serving in Vietnam. This was also authorized for service in Thailand, Laos, Cambodia, or their air space during the same, in indirect support of operations in Vietnam. Next to that is the Armed Forces Expeditionary Medal first established by Executive Order 10977, 04 December 1961, and qualified all U.S. military operations, U.S. operations indirect support of United Nations, and U.S. operations assistants from the four nations. The medal was awarded for service in the Republic of Vietnam from 01 July 1958 to 03 July 1965. Next to that is the Federal Service Medal. Worn above his ribbon bar is the first awarded Combat Infantryman Badge. To be eligible for the award of a combat infantryman's badge, individual must be an infantry officer in the grade of Col. or below, enlisted man, or warrant officer with an infantry military occupation specialty (MOS) and perform duties while assigned or attached to an infantry brigade, regiment, or small unit in actual ground combat. U.S. army personnel advising irregular forces, comparable with the infantry type units, such as the ARVN Rangers, and liaison personnel with the royal Thai Army and the Army of the Republic of Korea combat units in RVN, were eligible if they participated in actual ground combat with these units.[119]

Worn over his right shoulder is one of the newer items seen here, which are the rayon infantry blue shoulder cord, worn on his right shoulder, Infantry blue collar disks worn behind the U.S. and crossed rifles. Worn around his neck rayon bib-type scarf in the branch of service color (light blue) infantry, which could be worn beneath the collar of the shirt.[120] (These were first adopted during the Korean War, 1950-1953). On his right pocket flap is his name badge. Worn above that is the ARVN Ranger qualification badge. Worn on the epaulets are the green leadership tabs. E-7 rank is worn on each sleeve. Headgear is the maroon ARVN Ranger beret, with a gold bullion Ranger Badge.

U.S. ARMY ADVISOR, 38TH ARVN RANGERS, 1967

The ERDL camo tropical uniform is shown here in its short-lived cotton poplin version. The uniform was styled identically to the third pattern tropical combat uniform and included the same design features. Camouflage clothing was introduced to the army during WWII, and post-war studies of the spectral characteristics of commercial dyes made it possible to formulate both visual and infrared camouflage. The ERDL was technically restricted, being distributed first to pathfinders and reconnaissance combat units during 1968.[121] Although advisors like this one were able to bypass supply channels and obtain these camouflage uniforms through bartering with the supply sergeants, or by purchasing them on the black market. That same year, Co F 51st Infantry, a U.S. Army Long Range Patrol group, stated that the poplin material was too thin and ripped easily when wet. The LRPs also stated that the uniform could not withstand Vietnam's thick vegetation in which the long-range Patrols operated.[122] Photos show the entire advisory team of 36th ARVN Rangers wearing these poplin ERDL camouflage uniforms.

Two members of a Reconnaissance unit take a break during LRP training run by the Australian's Army Training Team at Van Kiep. The Ranger on the left wears tigerstripe with an ARVN Ranger pattern boonie hat, while the one on the right wears the more traditional ARVN Ranger pattern uniform with the maroon ARVN Ranger beret. Both wear Bata combat boots and have good luck bracelets.

Full front view of Co Van, 38th ARVN Rangers, in 1967.

Full front view of the advisor with emphasis on his left side.

Upper torso shot provides a good view of the advisor's sidearm, M1 carbine, and his Kodak camera. Note the grenade rings in his OD boonie hat.

Three Vietnamese Ranger scrolls from the 30th BN, 38th BN, and the 86th ARVN Ranger BN.

PAUL, NEED CAPTION, TWO BDQs, ONE IS 32 ON LEFT AND 38 ON RIGHT, AND OTHER IS 33 ON LEFT AND PB ON RIGHT.

Close-up showing advisor's ERDL poplin shirt with the full color B.D.Q. tiger shoulder sleeve insignia and 38th Vietnamese Ranger battalion scroll. Note also the Vietnamese made E-7 rank.

Three additional Vietnamese Ranger scrolls, featuring the 36th, 37th, and the 86th Vietnamese Rangers.

PAUL, TWO BDQs, LEFT HAS 6 AND 51, RIGHT HAS 5 AND 38, PLEASE PROVIDE CAPTION

Worn on both sleeves of the advisor's ERDL are locally-produced E-7 (Sgt. First Class) stripes, which are made out of OD cotton and black felt. Worn on his left sleeve is a Vietnamese-made Ranger scroll that reads 38th BN Vietnamese Rangers. The U.S. advisors only wore these. Below that is a silk woven BDQ Ranger SSI. Worn above his right pocket is a Vietnamese-made U.S. Army strip and Vietnamese made Airborne wings. Above his right pocket is a very crude, handsewn yellow ARVN Ranger qualification badge. Note the lack of a name strip. Headgear is an OD poplin boonie hat, with a single grenade pin in the foliage loops. Web gear consists of an M1944 OD pistol belt with a black leather holster for his .45. In his right hand is a Kodak instamatic camera. Weapon is a WWII era M-1 carbine with the long banana magazine.

ARVN RANGER ADVISOR, TRINH SAT, II CORPS, RECONDOS 1968

South Vietnamese reconnaissance instruction was an obligation of the Ranger Training command. Each one of the Division reconnaissance companies contained a Long Range Patrol Platoon. Military doctrine of the South Vietnamese high command em-

Full view of Ranger advisor attached to the ARVN Ranger Recon platoon, 1968.

Full view provides a good look at the ARVN Ranger's right side, and shows the AN/PRC-77 radio and the antenna and spare parts bag.

phasized observation only, and the use of the recon teams in a combat role was forbidden, and robbed the teams of flexibility, along with excluding valuable intelligence sources. For example, Combat Reconnaissance involving capture of prisoners was prohibited. Some recon units were never allowed to engage in patrolling duties and merely became command post guards. Recon teams and units were trained either at Duc My Ranger Training Centers' LRRP course, or at the Van Kiep National Training Center staffed by the Australians. The South Vietnamese Ranger cadre at Duc My taught a basic five week course for 150 ranger students. Emphasis was placed on physical fitness and the importance of reconnaissance techniques. Each session included three field exercises, which were always conducted in "safe areas." Every effort was made to avoid enemy contact, even though the Rangers were being prepared for combat.

The Australian Army Training Team began teaching a similar course at Van Kiep in June 1968. Their eight-week course was divided into two phases, and was intended for about 80 students. The Australians taught combat Recon tactics in addition to passive observation. Six weeks of formal instruction were given in ambush and patrolling techniques, followed up by two weeks of field combat exercises inside the Province of Phuoc Tuy.[123] Uniform consists of a set of heavyweight tigerstripe camouflage in

Upper torso shot shows the advisor utilizing the radio. Note the mixture of M1956 gear and the Korean War era pouch for the 30 round M-1 carbine. Note also the aluminum canteen in the M1956 canteen carrier.

Full view with emphasis on his left side. Note the BDQ tiger shoulder sleeve insignia.

Full rear view provides a good look at the AN/PRC-77 radio. Note the butt pack.

Close-up shows the right side of the Recon advisor and his web gear.

Close-up of ARVN Ranger Recondo badge, "Trinh Sat."

Upper torso shot provides a good look at the Ranger advisor's tigerstripe uniform, the BDQ SSI, and his U.S. made full color E-6 rank.

This provides a good look at the Ranger Recon advisor's tigerstripe uniform, not uncommon for the Recons.

the "silver" pattern. The shirt features two breast pockets, each with a billow on the inside, and is secured with a single plastic button. Worn on the right pocket is the II Corps Reconnaissance pocket insignia. On a black arrowhead bordered by yellow. Superimposed over the arrowhead is a black tab bordered in yellow with the inscription TRINH SAT (reconnaissance in red, hand sewn). Worn on the left shoulder is the Ranger shoulder insignia. Trousers feature two rear, two cargo, and a single "cigarette" pocket on the left leg, initially intended for a first aid dressing packet, which is sealed and contains a sterilized cotton gauze used for emergency dressing of wounds when medical services are not quickly available. Just above it in the cargo pocket are the LRP's notebook, pencil, and map, all in a plastic bag. Rucksack is a khaki tan Indigenous pack, which is based on an NVA 3 pocket sack and is constructed of heavy canvas. Each of its three outer pockets are capable of holding a single 1qt plastic canteen or other items of that size. Inside he carries his individual first aid kit, a poncho, half of a poncho liner, OD green towel, an AN/PRC battery, weapon cleaning kit, signal mirror, and enough rations for the duration of the mission.

Three Vietnamese Ranger scrolls from the 94th BN, 1st Recon, and the 93rd BN.

Close-up shows the advisor wearing the Recondo badge on the right pocket of his tigerstripe uniform.

III Corps Ranger group—this is a very unusual patch. Note the red "3" in the white box, and the red border; also note the length of the patch is much larger than the one on the right. The patch is silk woven, but is a very unique style, and very rare. The patch on the right is the III Corps 64th Ranger Battalion and is machine made.

III Corps Ranger signal tab features a maroon tab, bordered by black, with two white TT (Truyen Tin-Signal), and is attached to the insignia; machine sewn on Ranger camouflage. On the right is the III Corps Ranger Battalion, with a brown tab, and the battalion No. 31. Machine sewn on the ARVN Ranger camouflage pattern.

ARVN Ranger wearing the ARVN Ranger pattern, modeled after the U.S. ERDL pattern.

Four ARVN Rangers wearing the ARVN Ranger pattern with the BDQ patches and different colored maroon ARVN Ranger berets with different flashes.

U.S. RANGER ADVISOR, 52ND ARVN RANGERS Hoc Mon Canal, Plain of Reeds 1968

This Ranger Advisor to the 52nd ARVN Rangers at Hoc Mon Canal wears the U.S. ERDL uniform with numerous ARVN Ranger insignia. By late 1968 Ranger Advisors, as well as many other U.S. Advisors, were being outfitted with the ERDL uniforms. USMC and U.S. Navy Advisors were not as lucky and would have a harder time obtaining camouflage clothing. Many had to obtain them from U.S. sources, or wear the uniform of their host unit, such as the heavier ARVN Ranger pattern fatigues. On the advisor's left sleeve is an SSI of the ARVN Rangers, featuring the famous Black Tiger. Above that is an ARVN Ranger arc that reads "Biet Dong Quan," which is also silk woven and very rare. Above that is the U.S. Army RANGER tab, showing that Lieutenant Valentine had attended the U.S. Army Ranger School at Ft. Benning, Georgia, prior to being assigned to the ARVN Rangers. Worn above his left pocket is a hand-sewn U.S. ARMY tape, which is sewn onto a matching poplin ERDL strip

Close-up showing advisor's name tape, "Valentine," which, along with his hand sewn white and yellow BDQ qualification badge, are both sewn on poplin in E.R.D.L. camouflage material. Note the Vietnamese jump wings, which were added after his tour, but out of respect for LTC. Valentine, the wings were not removed.

Full figure shot of American advisor to the 52nd ARVN Rangers in the "Plain of Reeds," 1968. Note the black T-shirt.

Three quarter front right view of ARVN Ranger advisor. Note the MACV SSI that was added after the war by LTC. Valentine.

Full view gives a good shot of the advisor's WWII era Tommy gun.

Upper torso shot showing the hand sewn black and white silk woven Vietnamese wings, the 52nd Vietnamese Ranger scroll in full color, and the SSI of MACV. The MACV patch was also added after the advisor's tour. Note this shot gives a good look at the rank of Trung Uy (1st Lieutenant) on his maroon ARVN Ranger beret.

Upper torso shot showing advisor's left side provides a good view of the hand sewn "Branch of Service" BOS (Infantry) on his collar, along with his U.S. made Ranger tab, silk woven Biet Dong Quan Ranger arc, and silk woven Ranger shoulder sleeve insignia. Note also the Ranger's black T-shirt.

of material. Worn above that are U.S. parachute wings and a CIB. Both of these are also handsewn on poplin ERDL material. On his right side are handsewn nametapes "Valentine" on poplin ERDL material with a Vietnamese BDQ qualification badge. Worn on the pocket flap are Vietnamese Airborne wings, which were added after Lt. Valentine's tour of duty with the 52nd ARVN Rangers. On the advisor's collar points are the BOS (infantry) crossed rifles and his rank, 1st Lieutenant, which are both hand sewn onto poplin ERDL material. Lieutenant Colonel Valentine told me in an interview that these were all purchased at a large tailor shop in Saigon, where he had four sets made up for his assignment. Each patch cost about 25 cents. On the right sleeve is a Vietnamese-made Vietnamese Ranger arc of the 52nd ARVN Rangers, which were worn only by the U.S. Advisors. Below it is a U.S.-made MACV SSI, which was also added after Lt. Valentine returned home. The sleeves of the shirt have been cut short to make it easier to roll them up. Worn underneath is a black T-shirt.

Close-up showing advisor's U.S. Army, jump wings, and Combat Infantryman's Badge, which are also hand sewn onto poplin ERDL material.

Three Vietnamese Ranger scrolls.

Three ARVN silks BDQ Tigers. Note the bottom two are still attached and are quite worn.

One of the most distinguishing items of the ARVN Rangers was their colorful berets. The Biet Dong Quan berets were usually garnet in color, but varied sometimes, to orange, rose, or maroon. U.S. Army Advisors began wearing them officially in 1965, but many advisors had been wearing them as early as 1962. These berets were often bought a size too big and soaked in water and shaped into a "cock's comb." Worn on his head is another example of the BDQ beret. This one is more rose in color than some of the others, and also features the gold bullion "Arrow, Wing, and Shield" badge. Worn next to the ARVN Ranger badge is the advisor's rank, Trung Uy, or 1st Lieutenant. The third model jungle boots featuring the reinforced ankles and the Vibram soles are worn here. Weapon is the outdated WWII-era Tommy gun.

ARVN Ranger wearing a Vietnamese ERDL shirt with a subdued Ranger SSI.

Two ARVN rangers, both wearing the heavyweight Ranger pattern. Note only the Ranger on the left wears a BDQ Ranger badge.

Left to right: silk woven Vietnamese Ranger shoulder sleeve insignia. Features a yellow shield with a black border, and a large white star with a red border. Superimposed over the star is the Black Panther, headed with red features, a mouth and tongue, along with white whiskers and teeth. The patch on the right differs in that it is machine made on a yellow background with the maroon tab bordered in black, the battalion number 33 where 64 is usually, and white are attached to the insignia—these are machine sewn.

**U.S. RANGER ADVISOR,
33rd ARVN RANGERS,
BIEN HOA,
1969**

This Advisor to the 33rd ARVN Rangers wears another example of the ERDL produced in rip-stop cotton poplin. Above his left pocket is a very unusual U.S. ARMY tape, which has faded to almost white. Above that is a beautiful set of handsewn airborne wings, probably made in a tailor shop outside of Bien Hoa. Above the wings are worn a Vietnamese copy of the CIB that is machine made. Note, too, how it has faded to white due to inferior dyes and overwashing. On the left sleeve is a handsewn U.S. Army RANGER tab showing he is a graduate of the Ranger course. Below, worn as an SSI, is a silk woven BDQ panther head. This was usually worn on the right pocket, but some units wore it instead of the MACV patch. On the right sleeve is the Vietnamese

ARVN Ranger advisor to the 33rd Rangers, 1969.

Upper torso shot provides a good look at the advisor's ERDL shirt. Note the CIB on his shirt has faded to white.

Close-up showing advisor's U.S. made U.S. Army strip, hand sewn airborne wings, and a locally made Combat Infantryman's Badge.

Close-up showing the Advisor's patrol cap made from ERDL rip-stop material.

Close-up showing the silk woven B.D.Q. shoulder sleeve insignia, along with a hand sewn Ranger tab.

Close-up showing advisor's hand sewn 33rd Vietnamese Ranger battalion scroll, which is only worn by the Americans. This shot also provides a good view of his pin-on e-5 rank.

III Corps Ranger battalion features a No. 32 for 32nd Ranger battalion, and is also machine sewn on the Ranger pattern telephone material. To right is silk woven B.D.Q. shoulder sleeve insignia with a machine made 33 attached atop one piece, with that tab being added later in red.

From left to right, III Corps Ranger group, 64th Ranger battalion, and on the right is a 33rd Ranger Co. Both are machine made on the Ranger pattern camouflage material.

Ranger scroll in the traditional black and red. These were worn only by the American advisors and were worn on either shoulder. Worn on his collar points are the new-collar rank for enlisted and NCOs. This advisor's rank shows he is an E-5 Sergeant. Headgear is a locally-produced "patrol cap" in the ERDL rip-stop material. Once again, the local tailor shops could provide the American advisors with any style hat in any available pattern, and usually did.

U.S. RANGER ADVISOR,
51st ARVN Rangers,
Mekong Delta, VIETNAM
1968-1969

This heavily laden Ranger Advisor to the 51st ARVN Rangers wears the heavyweight, two pocket Ranger pattern fatigue shirt with no patches except the silk woven ARVN BDQ Ranger patch on his sleeve, typical of U.S. Advisors serving in 1969. The camouflage used by the ARVN Rangers and their U.S. Advisors was a direct descendent of the U.S. ERDL pattern. Starting in 1962, the Advanced Research Projects, Agency Research Development field unit (ARPA) began production and testing of a new reversible combat uniform for the Vietnamese. These test uniforms provided an ERDL leaf camouflage pattern on one side and black on the reverse side to resemble a pair of V.C. pajamas. By late 1964, the Biet Dong Quan had adopted the pattern, and it became known as the ARVN Ranger pattern, although other branches, such as the Airborne and the LLDB, wore and used this pattern. Although this pattern was recognized as the Ranger Pattern, some ARVN Ranger units continued to outfit their personnel in the Duckhunter pattern and Tigerstripe. The Ranger pattern was produced in a variety of colors and shades on different weights of material, but was still true to the original design. There has been some debate on which came first and where the variations were produced. CSM Mike Martin, author of Black Tigers, remembers the lightweight uniforms first appearing in 1965 while he was serving with the 44th. By late 1968, they resumed printing the distinct pattern on the heavier cotton material, although it is not clear if the production of the lightweight poplin uniforms ceased or continued in smaller quantities. By 1969, it seems the heavier weight Ranger pattern was predominant. The shirt features two small "patch pockets," reinforced elbows, shoulder reinforcements, and button cuffs. The matching heavyweight ARVN Ranger pattern trousers featured two front and two rear pockets only; later models would incorporate side cargo pockets much like those found on U.S.-manufactured jungle fatigues.

Many Advisors opted to carry their own radio, and this one carries the PRC-25. The AN/PRC 25 was partly transistorized, and depending on the terrain had an average range of about 3.5 miles, though greater distances could be reached with a 292 antenna.[124] The radio is seen carried here with the issue pack-board, which was very uncomfortable, because it kept the load against the wearer's back and prevented airflow. The radio handset is carried in the Advisor's pocket. Connected to the radio is a single M18 smoke grenade. This was the lifeline of the advisory team, and the fire support net for the entire Ranger battalion. On his M1966 vertical weave pistol belt he carries two Mk2 fragmentation grenades attached to his universal pouches, an M4 bayonet in an M8 scabbard for his carbine, a single first aid/compass pouch, two 1qt canteens with the M1956 covers, and a GI angle-head flashlight. Like many of his predecessors, he wears the garnet color Ranger beret with the bullion Ranger badge. The Ranger beret badge is made in bullion, and features a gold wreath, a sil-

Photo of unidentified ARVN Ranger advisor being awarded the Vietnamese Cross of Gallantry Medal in IV Corps. He wears the ERDL uniform with a full color RANGER tab and a subdued silk woven ARVN Ranger SSI. The Vietnamese officer wears the ARVN Ranger pattern fatigues and Ranger beret. Circa 1969-70. (Mike Martin Collection)

Heavily laden advisor from the 51st ARVN Rangers, 1969.

Ranger advisor wearing the ARVN Ranger beret takes a knee with the AN/PRC-77 radio.

III Corps. Ranger battalion features a battalion number. Both are machine sewn on the Ranger camouflage pattern.

Six Ranger group, with a blue tab bordered in black and a red No. 6 with a red bordered white square battalion number (35) on white. Machine sewn on ARVN Ranger camouflage material. On the right is a hand sewn BDQ patch, which is much larger than the average. The SSICHCV is hand sewn in white, and stands for "Command and Service (Chi Huy Cong Vu)." The patch features a maroon tab with light brown border, along with a hand sewn black border around the Tiger's head.

ver winged arrow, with a yellow shield, which features three vertical red stripes. Many American Advisors chose to wear these in lieu of the M-1 steel pot during operations. His M-1 Carbine is the WWII model, lacking the bayonet lug, along with a 30-round magazine and khaki strap.

U.S. RANGER ADVISOR, 38TH VIETNAMESE RANGERS, DA NANG, 1970

The Advisor has just been promoted to Captain and has had this new rank pinned to his beret in a ceremony, though he still wears his old rank of Trung Uy and 1st Lt. on his shirt. Although not a

From left to right are: the 32nd Ranger Group Administrative Service tab, which features a red 32 with white H.C.C.V., which stands for medicine and services; the tabs are connected to the patch, which is machine made on the ARVN Ranger camouflage pattern. To the right is the III Corps six Ranger group, which features a red No. 6 in a white box, and blue tab with a white No. 35. Note the difference between the two. The patch features yellow whiskers on the tiger head.

ARVN Ranger advisor to the 38th ARVN Rangers.

Another shot shows the 38th ARVN Ranger advisor wearing the famous BDQ "Black Tiger" painted on the M1 helmet, along with his outdated Tommy gun.

III Corps Ranger headquarters features a blank maroon tab with no numbers, with a black border on the ARVN Ranger pattern.

common practice, some Advisors chose to wear their U.S. rank on their berets as opposed to their Vietnamese equivalent.

His ERDL uniform is U.S. made and adorned with both American and Vietnamese patches, typical of Advisors in 1969-1970. On his sleeve he wears a subdued twill MACV patch. Above that is a black and red, hand sewn Vietnamese Ranger scroll of the 38th Vietnamese Rangers worn only by American personnel, along with the black and gold tab shows he is a graduate of the American Ranger school. Centered on his shirt is his Vietnamese equivalent, Trung Uy. On his pocket is the silk woven ARVN Ranger BDQ patch. Above his name strip is another example of a hand sewn Ranger Qualification patch. Note his military Timex watch through his buttonhole, and black T-shirt.

Same Advisor, wearing the brightly painted Ranger helmet which is typical of the helmets in Ranger units. Both the Vietnamese wore these and their Co Vans while on operations in "Indian Country." The helmets were painted by paying local shop owners, by the wives of the Rangers, or by finding talented men

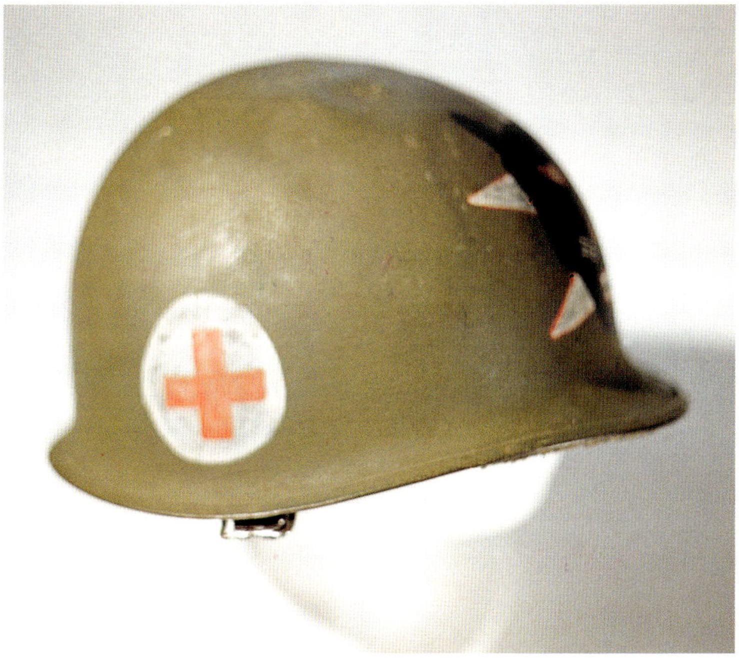

Close-up of red cross on painted Ranger helmet. This would indicate it belongs to the medic.

Close-up of giant "Black Tiger," sometimes called a panther head, painted on the ARVN Ranger helmet.

Additional ARVN Ranger scrolls from the 22nd, 23rd, and 30th BN.

Three ARVN Ranger scrolls from the 32nd, 33rd, and 31st Ranger BN.

in the unit and tasking them with the chore. Each unit would paint the fierce Ranger tiger/panther, a larger version of the BDQ patch, on a different background. For example, the 44th had a yellow and black paint scheme, while the 32nd Ranger Battalion chose bright green and brown for its background. His weapon is a WWII-era Tommy gun.

Two Vietnamese made Ranger SSIs. both are machine made on ARVN Ranger material.

5

U.S. Navy Advisor 1963-1972

BRIEF HISTORY: JUNK FORCE

Law has defined the mission of the Vietnamese Navy as follows: provide for security of sea approaches of South Vietnam, including patrols, escort, mine sweeping, mine laying, anti-submarine warfare, and in wartime, defense of designated harbors.

Provide for the security of rivers and canals in the Delta area, including patrols, escort, mine sweeping, and, in wartime, defense of designated river ports.

Maintain ocean and river amphibious capabilities and conduct joint operations with the Army and Air Force when directed by the joint staff.

Maintain the Marine Corps in combat readiness, trained in amphibious warfare to conduct separate operations and/or to support and to reinforce the Army forces.

Provide direct fire support to the ground forces.

Provide water transportation as available in the Vietnamese Navy: ship and boat units, both coastal and inland, for personnel and materiel of the armed forces of Vietnam.

Maintain naval stations and sea and inland water patrols to assist customs and security force authorities as prescribed by the joint staff.

Provide emergency assistance for, and, in wartime, protect the Vietnamese Merchant Marine ships at sea, assist in the administration of the ocean fishing laws of Vietnam.

Organizations; The Vietnamese Navy is composed of the Naval staff and the Naval establishment.

The Vietnamese Navy staff is headed by the Commander of the Vietnamese Navy, who was also the deputy chief of staff for the Navy on the ARVN General staff. A staff comprising a Chief of Staff and four Staff Divisions assist him as follows:

The General organization and personnel (N-1)

Operations, including intelligence, training and communications (N-3)

Logistics (N-4)

Psychological War (N-5)

The Inspector General of the Navy is responsible directly to the CNO for judging the combat potential of the Naval units and auditing the use of the Navy funds.

The Vietnamese Naval establishment is composed of the operating forces and the shore establishment.

The operating forces consist of three major commands:

The *Sea Forces*, which are composed of various small type ships and craft including PC, PCE, LSM, LCU, LSIL, LSSL, and MSO, which are grouped into FLOATILAS by ship type. The sea forces General mission consists of patrolling the 2000 KM the South Vietnam coastline; conducting coastal surveillance operations and anti-Viet Cong patrols; providing amphibious lift fire support; harbor defense and mine sweeping; providing administrative the logistic lift for the various government agencies.

The *River Forces*, which consist of LCU, LCM, LCVP, and river patrol boats of French design, are heavily armed and armored. These forces are committed to continuous operations in the Mekong Delta area under operational control of the Third Corps Commander. The General mission of the river forces is to provide tactical troop lift, fire support, and logistic water lift for ground forces to carry out patrols, escort, and convoy duties in the canals, rivers, and waterways of the Mekong Delta region.

The shore establishment consists of the following:

The Naval Headquarters is located in Saigon, which includes the office of the Vietnamese CNO , his staff, and intelligence, administrative and communications elements of the Vietnamese Naval organization.

The stations and training command, comprising:

Five Naval Districts located at Saigon, Danang, Nha Trang, Vung Tau, and Phu Quoc, located in the Gulf of Siam.

Headquarters company, including Bach Dang Camp, military police, and the Navy band.

Naval training center at Nha Trang, the main Vietnamese training command, including a midshipmen academy, class "A" and "B" schools for Petty officers, and training facilities for new Naval recruits.

Underway training center is located in Saigon, and charged with underway training of the various ships and boat crews.

The River Force Bases, established at Saigon, Long Xuyen, Vinh Long, My Tho, and Can Tho which provide local operating bases for the small boats of the River Forces.

The western repair facilities at Can Tho, the major maintenance base of the River Forces. A Marine Railway, near completion (1961), will eventually improve the capability of the WRF to repair boats of the river forces. Due to present limited facilities, many river force boats must go to the Saigon naval shipyard for repair and overall.

The Naval Shipyard, located Saigon, which provides overall, maintenance, and technical repair services to the Vietnamese Navy. The shipyard is the largest of its kind between Singapore and the Philippine islands.

In the Naval supply center, located in Saigon, which provides supply support for the Sea Forces, River Forces, and the shore establishment of the Vietnamese Navy. This activity requisitions and stores supplies, conducts inventory, compiles usage control data, and trains Vietnamese personnel in supplies procedures.

The Naval Administrative Center, located Saigon, is responsible for disbursing.

The Secretariat of Navy command, located Saigon, is charged with handling CNO correspondence.

The Medical Corps. is in charge of organization and control of medical matters within the Navy.

The Junk Force was organized by a presidential decree as a paramilitary force and as an inshore Patrol Force. The Junk Force has been under the personal command of the Naval Deputy and is manned in part by naval personnel, with Vietnamese fisherman providing the bulk of the force.

In general, the Vietnamese Navy, while small in size and consisting of relatively inexperienced personnel, is slowly developing into an efficient force capable of performing its mission (1961).

"According to Geneva agreement, concluded after Dien Bien Phu on 20 July 1954, the French are to remain in the southern half of the now divided Vietnam until the general elections, to take place in July 1956. An effort to synchronize American and French advisory functions in Vietnam, a training relations instruction mission was created pre-December 1954 consisting of one Navy, one Air Force, and three army officers from each nation. The mission grew rapidly, having 33 U.S. and 67 French officers, but American naval representation remained low. By May 1955, the U.S. contingent had reached 155, yet there were only two naval officers in the mission."[125]

"Command of the Vietnamese Navy had been turned over to the Vietnamese in July 1955. The French command in Vietnam was disestablished on 26 April 1956, and two days later the training relations instruction mission was closed down.

"Almost immediately upon taking office, President Kennedy began grappling with the situation in Indochina. First Laos, then Vietnam became important considerations of his administration, although events in other parts of world, such as Cuba and Berlin, continually competed for his attention. During his first year he sent his vice president, Lyndon Johnson, to Saigon, the capital city of South Vietnam, to cover president Diem. The discussion between Johnson and Diem centered on the enlargement of the U.S. commitment to South Vietnam and included some specific talk about strengthening the Vietnamese Coastal Patrol Force."[126]

"For quite some time communists had been suspected of bringing supplies into South Vietnam from the sea. The Vietnamese Navy had begun patrolling the coastal waters in junks, the traditional Oriental sailing vessels, and it was this junk force, as it was called, that would need considerable expansion in order to have any significant success in interdicting the flow of communist supplies. At a National Security Council meeting on 29 April 1961, President Kennedy approved the training of Vietnamese junk crews in Vietnam or at U.S. bases by U.S. Navy personnel."[127]

"Vice President Johnson's proposal to strengthen the junk force was endorsed and expanded, to include controlling the inland waterways, as well, by General Maxwell Taylor, President Kennedy's personal military advisor, who had gone to Vietnam in October at the President's request to assess the military situation there. Taylor sent a cable gram to the President proposing extended expansion to the U.S. commitment, in which he said 'the U.S. Government will assist the GVN in effecting surveillance and control over the coastal waters and inland waterways, furnishing such advisors, operating personnel, and small craft as may be necessary for quick and effective operation.'"[128]

"On 11 November 1961, Secretary of State Dean Rusk and the Secretary of Defense Robert McNamara sent a joint memorandum to the President outlining a series of recommendations and priorities for coping with the Vietnam situation. Among the specific proposals listed as immediate in nature was a call, similar in wording to General Taylor's message, for providing the small craft, 'uniformed advisors and operating personnel' as necessary to exercise control over the coastal waters in inland waterways." [129]

Yabuta Class Vietnamese Motorized Command Junk on Market Time patrol in the Gulf of Thailand. You can see the .50 cal on the bow and the .30 cal on the stern. During river ops. a mortar was mounted in the center where the advisor is standing. (Sam L. Pfiester Collection)

"To respond to these recommendations from his top advisors, President Kennedy significantly enlarged the U.S. military effort in Vietnam. Although stopping short of sending combat troops, which had also been repeatedly recommended to him, he substantially increased the number of American advisors over the three years of his administration. By the end of 1961, the number had quadrupled to exceed 3000; by mid 1962, there were 12,000; and by the end of 1963 the total passed 16,000. While most of these additional advisors were not naval personnel, President Kennedy, himself a naval officer in WWII, did not overlook the importance of a naval role in Vietnam. Vice President Lyndon Johnson wrote of 1961 in his memoirs: 'we increased assistance to the Vietnamese Navy to enable it to protect the coast against infiltration from the North and to patrol the inland waterways used extensively by the Viet Cong.''[130]

"With this increased support from the United States, the Vietnamese Navy grew by leaps and bounds between 1961 and 1965. The 3 task oriented forces had been established: the Sea Force, the River Force, and the Junk Force. The Sea Force, which consisted of the larger, more conventional coastal craft, grew from 21 to 48 ships during this period. The River Force, designed primarily for transporting troops along the waterways, increased its assets from five River Assault Groups, consisting of approximately 100 boats, to 7 RAGs with a total inventory in excess of 150 boats. The Junk Force grew to 644 craft by 1965. Total Vietnamese personnel expanded from 3000 in 1961 to more than 8000 in 1965."[131]

"The number of American naval advisors grew correspondingly from 53 in the MAAG in 1961 to 235 in the Naval Advisory Group (NAG) component of MACV by the beginning of 1965. (MAAG was absorbed into the newly formed MACV—Military Assistance Command Vietnam—in May 1965) As the number of advisors increased, more began to advise at the unit level. Increasingly, advisors appeared on Sea force ships, with River Assault Groups, and aboard junks. The Die was cast: the American Navy had begun its coastal and riverine involvement in Vietnam."[132]

"By this time American advisors were living side-by-side with their Vietnamese counterparts. Some were assigned to the ships of the Vietnamese Navy sea going forces. Others were patrolling the rivers of the Mekong Delta in converted WWII landing craft as advisors to the River assault group. Some lived at remote coastal bases in thatched huts with members of the Vietnamese junk force. Still others were helping to develop the curricula at the naval school in Nha Trang, and a substantial number were posted in Saigon to assist the shipyard or at the Vietnamese Navy headquarters."[133]

When America had first become involved in Southeast Asia, no Coastal or Riverine patrol craft had been in the Navy's inventory, no official doctrine existed for either kind of warfare, and even the unofficial ledger appearing in Navy oriented periodicals and journals reflected a paucity of interest in these operations. A study of American naval history would tell him that there was certainly plenty of precedent—coastal and Riverine warfare had played significant roles in the revolution, the War of 1812, the Mexican War, the Seminole and Creek Wars in the Florida Everglades (1835 to 42), the Civil War, and the Yangtze river patrols in pre-WWII China, but record was not a continuous one. "Brown water warfare" had been used time and again, but on every occasion, once a necessity had passed, these capabilities were shelved and the Navy returned to "blue water" operations. The only connection between small craft operations conducted in the past and these modern-day sailors was a history book.[134]

"Newly assigned Naval Advisor Capt. Hardcastle, himself a veteran of combat action in WWII and Korea, respected these young men, but he also worried that in their zeal, caution might be overlooked. He did not want to lose any of these fine officers. Because the dedicated man often neglected their health in their quest for achievement, they were all encouraged and occasionally ordered to go back to Saigon for a breather and for much-needed medical attention. The primitive living conditions in Hilo were without the conveniences that most Americans take for granted, such as running water and sewage disposal, and resulted in cases of tape worm, diarrhea, dengue fever, and other maladies. Many advisors suffered severe weight loss during their time in the field. Capt. Hardcastle visited every location where he had advisors at least once a month to monitor their conditions, noting that because of their conscientiousness, these men would keep putting off their recuperative trips to Saigon. There was always one more thing to do."[135]

"One of the issues raised by Secretary of Defense Robert McNamara upon his return to Washington from the Pacific in early 1964, was the possibility of streamlining the U.S. commanded organization by integrating the Military Assistance Advisory Group into the Military Assistance Command Vietnam, which already was charged with coordinating the total American effort in Vietnam and advising senior South Vietnamese military and political authorities. Admiral Felt opposed this reorganization. He believed that the MACV commander could not effectively control the large organization which would result from the merger of the two commands. Further, the existing setup allowed MACV to concentrate on the counterinsurgency effort without being bogged down with details of the military assistance program and the direction of the growing number of field advisors. The Admiral reported to the joint Chiefs that, 'we will be unduly rocking the boat to no practical purpose since COMUSMACV already clearly exercises operational command over MAAG and advisors."[136]

"Naval Advisor Capt. Drachnik favored the reorganization, based on his observation of several problems resulting from the divided command. He noted that although operational control in the American effort in Vietnam rested with MACV, MAAG allocated the military systems' program supplies that supported operations. Since operations obviously depended on supply, he argued that the two functions should not be separate. Capt. Drachnik also concluded that since the MACV Commander only advised high-level Vietnamese officers, he cautioned with little feel for the problems of MAAG's field advisors. The joint Chiefs submitted a split opinion to Secretary of Defense, with the Army favoring the consolidation and the Navy opposing it. Admiral Felt continued to doubt the need for the reorganization. Nevertheless, on 8 April McNamara approved the merger. The MAAG was disestablished on 15 May. At the same time, the Navy section of the MAAG was redesignated the Naval Advisory Group, MACV, under Capt. William H. Hardcastle, who relieved Capt. Drachnik in January 1964. As a result of the reorganization, the number of men authorized for MACV increased slightly from 3,369 to 3,677. At the beginning of the year the number of naval advisor in Vietnam numbered 154. But these figures soon increased substantially. When the Kennedy policy of phasing down U.S. involvement in Vietnam was set aside in 1964, the JCS authorized an increase of MACV advisory personnel. By the end of 1964, they numbered 4,889, 235 of whom were from the Navy. Additional supporting units attached to MACV brought the total of U.S. military forces to over 23,000 at that time"[137]

"The 81 additional naval advisors who reached South Vietnam in 1964, were assigned primarily to field billets, although psychological warfare personnel and medical advisors also were added at the headquarters level. The 21 new Billets for the Sea Force allowed advisors to ride individual ships on a regular basis in order to promote more aggressive patrolling. Advisors of the River Force increased from ten to 32, which enabled the establishment of three man detachments, with each of the six existing RAGs, with a new SL group, and with the river transport escort group. The total to advisor's age were assigned to the eastern and western repair facilities. The remaining personnel staffed administrative positions. The assignment of additional advisors to the Coastal Force in 1964, enabled the stationing of two men at each of the junk repair facilities at Danang, Nha Trang, Cat Lo, and Rach Gia, and three with each of the 8 advisory detachment assigned to the Coastal Districts. Besides instructing Vietnamese on ways to improve naval operations, the advisors were now able to provide Capt. Hardincastle and the Military Assistance Command in Saigon with more information on the true situation in the field."[138]

"The increase in advisors partly reflected the growth of the Vietnamese Armed Forces, the expanded support of which was authorized after secretary McNamara's visit to Vietnam on March 1964. The number of Vietnamese Navy personnel rose from 6,467 in January to 8,162 officers and men in November. Gen. Khanh's national public service decree, signed in April 1964, which made all men between the ages of 20 and 45 subject to the draft, aided the Navy's recruiting program, although the services were manned solely by volunteers. Additional personnel manned the ships and craft arriving in the United States, raised the manning level of other units and repair facilities, and filled the training pipeline."[139]

Rank Insignia of Navy Enlisted People (E)

Paygrade/Rank	Hat and collar	Sleeve
E-1 Seaman Recruit		
E-2 Seaman Apprentice		
E-3 Seaman		
E-4 Petty Officer Third Class		
E-5 Petty Officer Second Class		
E-6 Petty Officer First Class		
E-7 Chief Petty Officer		
E-8 Senior Chief Petty Officer		
E-9 Master Chief Petty Officer		
E-9 Master Chief Petty Officer Of The Navy		

Rank Insignia of Navy Commissioned Officers (O)

Paygrade/Rank	Caps, shoulder and collar	Shoulder "boards"	Sleeve
O-1 Ensign	gold		
O-2 Lieutenant Junior Grade	silver		
O-3 Lieutenant	silver		
O-4 Lieutenant Commander	gold		
O-5 Commander	silver		
O-6 Captain	silver		
O-7 Rear Admiral (Lower Half)	silver		
O-8 Rear Admiral (Upper Half)	silver		
O-9 Vice Admiral	silver		
O-10 Admiral	silver		
O-11 Fleet Admiral	silver		

BRIEF U.S. NAVY ADVISOR UNIFORM HISTORY

The first U.S. Naval Advisors arriving in Vietnam wore the standard tan class B, summer tropical, with overseas cover or Khaki tan visor cap. The khaki shirt would be worn with rank device, nametag, and the Vietnamese junk force device worn on the right pocket on a leather strip as a pocket hanger. Aboard the boats, photos would indicate cotton sateens with only rank, and no U.S. Navy or nametape, and the officer or enlisted man's visor cap, khaki. Later, in early 1963-64, as more Army Advisors were being issued the new Tropical Jungle fatigues, the Navy was slowly getting hold of pieces and complete uniforms, too. These would be worn in accordance to Army doctrine, with a U.S. Navy tape and nametape, in yellow on blue or white on blue, depending. Pin-on rank would be worn on the collar. On the right sleeve would be worn the MAAG Vietnam patch in accordance to Army regulations. Also, later many Navy personnel would wear the MACV patch. Headgear consisted of Vietnamese made "cowboy" hat, which featured a wide brim and could be bought anywhere in Vietnam, or a dark blue Navy style baseball cap, similar to the Army's but dating back to pre-WWII. These have been seen with rank devices and or junk force or RAG force devices affixed to them. Other photos show some advisors wearing the standard Navy cover, or even Aussi bush hats. Black berets with the Junk Force emblem and the NAG device were worn also, along with a whole host of other berets in every pattern imaginable were worn with either device affixed. Bruce McIver, R.A.I.D. 72, 1969, wore a tigerstripe boonie on occasion with the NAG beret flash affixed to the front. Another oddity worn by Bruce was the NAG beret flash as a pocket device, on his left pocket. Uniforms consisted of standard cotton sateens, Jungle fatigues, tigerstripe uniforms, and, in one instance, a set of ARVN Ranger pattern worn by Don Slinkard with name, U.S. Navy, and rank all direct embroidered in yellow. Also, many naval advisors keeping with local and paramilitary customs wore the uniform of the junk force, and their uniform was peasant garb, which was both cheap and functional. Flak vests were worn by both Army and Marine Corps. These were worn only while on the boats, along with an M-1 helmet. Weapons varied from M-1 carbines and M-16, M-79, Tommy guns, and German MP-40 machine pistols, to the occasional MAT-49. Side arms ranged from standard issue 38. Cal and 45. Automatics to 9 mm Browning, German Lugers, to captured Chi Com pieces. Vietnamese Naval personnel wore everything from the black PJs of the Junk Force, OD green cotton sateen uniforms, and even light blue denim shirt and dark blue denim trousers, "dungarees" working uniform with a dark blue "Dixie cup." Photos show headgear worn by the Vietnamese were Army style baseball caps with the Vietnamese Naval Device sewn on or direct embroidered, visor caps with the same device, or the Vietnamese blue Naval beret. It seems there were no set regulations, and headgear was different depending on what Province you were in.

JUNK FORCE ADVISOR; JUNK DIVISION PHU QUOC; 4TH DISTRICT FLEET; CHIEF PETTY OFFICER; 1964

In 1964, American Advisors in Vietnam numbered 16,000. Of these, a small number were U.S. Navy personnel advising the "Junkies," the junk force sailors. These Vietnamese sailors, nicknamed the "black pajama navy," were part of the para-military at the time, and not part of the Vietnamese armed forces. Their mission: to patrol the coast of Vietnam, along with the "Biet Hai," which were the Vietnamese Commando groups attached for raids ashore, along with stopping the flow of communist supplies from the North and interdicting the movement of the V.C. Cadre. This "ragtag" fleet, the "Hai Thuyen," or junk force, was first created in 1960. The American Advisory Team, which usually consisted of 3 men—one Lt. and two Chief Petty Officers—would ride on the command junk, called the "Chu Lucs." Unfortunately, many of the junks were lost or damaged caused by wood boring torpedo worms, which drilled their way into the hulls of the junks and make them look like Swiss cheese. More junks would be lost in one year to worms than to combat. New and better junks were introduced two years later, called "Yabuta" class, after the Japanese Technician, along with the older junks being updated and outfitted with a fiberglass coating to keep the worms out.

Chief Petty Officer, Fourth Coastal District, Chief Naval Base, Phu Quoc, February 1963.

Full figure shot provides a good view of the Chief's BAR machine gun.

Uniform: Tattoos, and black pajamas. This chief is one of the four members of the U.S. Navy who were stationed on the island Phu Quoc as advisors to a ragtag fleet of 140-armed junks. The clumsy wooden crafts belong to the black pajama navy, which patrols some 1,000 miles of coastline from Cambodian border all along the South China seacoast to the 17th parallel. The Chief and the three others live in a crude shack in the village of An Thoi where they share rustic living quarters. An Thoi is the site of an old French Penal Colony, and its crumbling cells now serve as sleeping quarters for sailors of the Fourth District Fleet. Like their sailing craft, sailors in the black pajama navy are an unlikely assortment. Many of them were hastily and sometimes forcefully recruited off the streets of villages and cities in South Vietnam. Each of the 4 district fleets has approximately 140 junks. Weather, particularly during monsoon season, can be as devastating as the V.C., and cost the S. Vietnamese Navy 50 junks in 1963. In 1960, the U.S. suggested to the South Vietnamese that they form a wooden naval force as a means of denying the long coastline to infiltration by the Communists.

Close-up of the Chief's visor cap, worn during operations on board the rag tag Junk fleet.

This re-creation of a Chief Petty Officer wears the typical black cotton pajamas, same as the V.C. and local peasants. The "uniform was light weight, loose fitting, and well suited for hours on board the Vietnamese Junks. Many U.S. Navy Advisors wore these along with standard Navy/Army cotton OG 107 sateen, sometimes inter-mixing the two in a hog-pog of quasi-military uniform. Some higher Fleet officers disapproved of the wearing of any South Vietnamese Navy uniform, and/or para-military uniform, to include the civilian black PJs. Headgear worn here is the Navy's Petty Officer tan service cap with visor. Most junk force personnel and their U.S. Advisors wore the black junk force beret with junk force emblem, which was a twin sail junk device the size of a silver dollar, silver for enlisted and gold for officers. In his hand he carries a WWII era Browning Automatic Rifle (B.A.R.), which, along with a second one and an M1919 30. cal. Machine gun, would be used to arm the junks. Many of the "Junkies," or junk force sailors, were armed with M-1 carbines, Thompson sub-machine guns, M-3 grease guns, German MP-40s, and M-1 Garands. This advisor's personnel side arm, a snub nose Smith & Wesson 38. Cal pistol, is kept secure in the pocket of his pajama trousers. Boots worn here are the early Black Combat boot, sometimes mistaken for Jump boots due to the toecap. During 1962, the black leather combat boot of a new design was adopted as part of the "standardization" of supply throughout the armed forces.

U.S. NAVAL ADVISOR;
R.A.G. 23,
3d Coastal District
Mang Thu River,
1964

"River Assault Group 23 is one of the Vietnamese Navy's most daring gunboat forces. Each group, known for its initials as RAG, operates a miniature fleet of shallow draft patrol craft. A Vietnamese Lieutenant and advisor command each by an American naval officer. The RAG forces constitute the military presence of South Vietnam on thousands of miles of waterway spanning the vast Delta region of Indochina's Mekong River."[140]

"The Vietnamese Navy's so-called junk fleet patrolled the coasts of the South China Sea and the Gulf of Siam. The river gunboat groups, the R.A.G.s, operated across the broad flood region between the two coasts. RAG 23 was responsible for safeguarding three of the 13 Delta provinces—Vihn Long, Vinh Binh, and Kien Hoa. These are crisscrossed everywhere by majestic rivers, meandering streams, and straight canals. The way of travel and even the way of life here is water borne, aboard sampans and junks. Most people rarely see or used the few highways of Vietnam. Rag 23's home base is the provincial Capital of Vinh Long, and more than 75 miles southwest of Saigon. There are a total of

U.S. Navy advisor (NAG) to the Vietnamese River Assault Group (RAG) 23, operating base in Ben Long in the Mekong Delta, 1964.

Another view shows the Naval advisor with his OD green "cowboy" hat and British Sten sub-machine gun.

seven assault groups, each with 19 boats of various types, which combat the Viet Cong in thousands of miles of river and canal lacing the vast rice bowl known as the Mekong Delta."[141]

"The Advisor and his crew have swept down the Mang Thu River the previous night to aid a battled district headquarters, Cai Nhum. The Viet Cong had broken off their assault as they arrived. The Viet Cong are shifting the priority target from Vietnamese to Americans."[142]

"Vinh Long province, like the rest of the Delta, had never been truly a part of any Indochinese state. Here Khmer control ebbed centuries ago. The French, a generation back, sometimes taxed, but never ruled. Before River group 43 appeared, many villagers had not seen a representative of Saigon since the 1950s. Local Warlords, River Pirates, and now the Viet Cong had installed themselves without hindrance from any central government."[143]

R.A.G. and its U.S. Advisors, usually 3 men, two of whom were Chief Petty Officers and this recreated Officer. R.A.G were responsible for 3 of the 13 Provinces, Vinh Long, Vinh Binh, and Kien Hoa. The RAG home base is at Vinh Long, which is 75 miles South West of Saigon. At this time there are seven Assault Groups, each with 19 boats of various types, to combat the V.C. in thousands of miles of river and canals in the Mekong. This U.S. Advisor goes by the call sign, "Horn Blower," from the spirited "hero" of C.S. Forester's novels, and demonstrate their connection to the past by flying the "Jolly Roger" flag on the command junk. Their loyal Vietnamese sailors have taken their cue from these salty Americans and begun calling themselves "the River Rats."

This Advisor rides on the 60 foot Command Gunboat, which is lightly armed with a turret mounted .50 cal. and three .30 caliber WWII machineguns.

Uniforms worn by the members of the U.S. Navy Advisory team are the first model "exposed" button jungle fatigues, usually worn with the sleeves rolled up. Many of the Navy Advisors had a U.S. Navy stencil on the left pocket, or chose matching blue and white name strip and Navy tape. Worn on the left sleeve would be the U.S. Army MAAG patch. Headgear worn here is a typical Vietnamese made "cowboy" hat in OD green. These hats featured two aluminum screened eyelets on each side, to help keep the wearer cool, a long chin strap, and a wide brim to keep the sun off the advisor's face, neck, and ears. This could be bought at any of the local tailor shops in or around Saigon. These hats can trace their birth back to the French involvement and the wide brim hats they brought with them. Jungle boots are the second model Tropical Combat Boot, which did away with the leather strip along the top and rear of the boot. 1965 would introduce the third model with the nylon reinforcement at the ankle. Weapon is the WWII Mk II Sten sub-machine gun in 9mm with 30 round magazine, which was not uncommon to the RAG sailors and their American counterparts.

U.S. NAVAL ADVISOR; COASTAL GROUP 23, "JUNK FORCE" SONG CAU 1965-66

"The Vietnamese Navy had little trouble recruiting for the regular forces, but meeting the allowance for the paramilitary coastal force remained typical. As of 01 January 1964, the coastal force was authorized 6,640 men. But the force never had more than 3,700 men onboard during the year. Desertions, which since April 1963 averaged about 100 men each month, were the chief problem. Between May and August 1964, only 264 men were enlisted to offset this loss. Under manning was most serious in the Third and Fourth Coastal Districts. Coastal Division 34, for example, had only 40 junkmen of the 145 authorized actually onboard. At the same time as recruiting fell behind, almost one-third of the coastal force personnel were assigned to duties other than the primary one of coastal patrol. Thus, the number of junks on patrol hovered around 40 percent of the total operational force. As a long-term solution to the personnel shortage, U.S. advisors called for the formal integration of the junk force into the Vietnamese Navy. This would make personnel eligible for all the benefits of the regular service, including higher pay and allowances, and thus enhance the desirability of coastal force service. Although the Joint General Staff expressed concern that the actual increase in personnel costs set an undesired precedent for other paramilitary units in the armed forces, and it approved the integration plan in December. During the following summer, the coastal force formally joined the Vietnamese Navy."[144]

"The junk divisions' function was simply to check every coastal boat that comes or goes, and to be sure it is not serving the Viet Cong. The junk sailors and their advisors try not to do fighting on land. Although, the sailors have made several land raids, but only to secure their own bases. Most junk bases lie at the edge of an area traditionally under Viet Cong control. Within a day's walk of where the boats are moored there are said to be deeply entrenched Viet Cong that operate a rest camp, a hospital, and officers' training center."[145] "A training program, set to handle rapid expansion to the Navy in 1962 and 1963, began to function

Full view of Junk Force advisor, Junk Force Base, Coastal Group 23, Song Cau, 1965-66.

Full view of junk force advisor showing his U.S. made black leather shoulder holster for his M1911A1 .45 cal pistol, along with his "official uniform," black PJs.

Full view of junk force advisor, geared up for possible action aboard the Command junk. This gives a good view of the silver non-commissioned officer's junk force badge affixed to his black beret.

routinely by 1964. With the graduation of the Naval academy class in October 1964, the previous Officer shortage ended. An 8-week tour of duty with 7th Fleet ships was provided to 25 percent of each class. To give refresher training to Vietnamese technicians in engineering, deck, and gunnery procedures, and operational skills, U.S. advisors helped the Vietnamese set up 35 courses, varying in length from three days to four weeks, at the Naval Advanced Training Center in Saigon."[146]

"The prototype junk selected for construction and eventual use by members of the 23rd Coastal Group was a 57 foot boat with a fiber-glassed wooden hull, one of which had been built in 1961 from the design of a Japanese engineer, Mr. Yabuta, at the Saigon Naval shipyard. CINCPAC approved the findings for 90 of these craft in the fiscal year 1965 military assistance program budget, and Admiral K. enthusiastically approved construction of the craft in the Saigon shipyard. The program began in Octobe 1964, with all of the boats scheduled for completion within one year. Capt. Hardcastle later cited the Yabuta program as one of the outstanding achievements of the advisory effort."[147] This U.S. Naval Advisor is one of the first to receive one of these new Yabuta Class junks at his base.

"By the beginning of 1965, there were 81 command junks, 91 combination motor/sail junks, 120 motorized junks, and 239 sailing junks distributed among 28 bases on South Vietnam's coast. An addition to the Yabuta class, Capt. Hardcastle gave consideration to obtaining a steel-hulled craft requiring less maintenance and offering a longer useful life than the wooden junks. The chief of the Advanced Research Projects Agency's field unit in Vietnam was sympathetic to Capt. Hardcastle's request for an evaluation of such a vessel.

He commented:

"The concept of the junk force was a stop gap quick fix at best. At a time when the Army of Vietnam is using M113 armored personnel carriers and the Vietnamese Air Force is using A1H Sky Raiders, the Navy of Vietnam is attempting to patrol its coast with craft which might have held their own in the battle of Salamis."[148]

This Chief Petty Officer wears another example of the traditional black pajamas utilized by both the junkies and U.S. advisors alike. The pajama top features a button closure, an open collar, two open lower pockets, and a single open breast pocket on his left breast, just large enough for a pack of cigarettes and a Zippo lighter. The shirt was designed to be worn outside the matching black trouser, which featured an elastic waistband and two front slash pockets. These black PJs were lightweight and cool, and weighed about half as much as the U.S. Army jungle fatigues and one third less than the uncomfortable U.S. Navy issue OD cotton sateen uniform. Some advisors wore a combination of black pajamas and U.S. in a type of quasi-military uniform. 2nd model jungle boots are worn here, although some junk advisors opted to wear tennis shoes, flip-flops, or no shoes at all. Worn across his chest is a black leather M7 shoulder holster for his Colt .45 M1911A1 pistol. Additional magazines could be carried in the two shirt, or in his trouser pockets. Headgear is a Vietnamese made black beret with the silver twin sail NCO junk force device pinned to it.

Worn by the same junk force advisor, geared up for action onboard the Command junk, is the Marine Corps M1955 Armor-Body, Fragmentation Protective, commonly known as the flak vest. This was an updated and improved version of the M1951 vest that was developed and tested during the Korean War. Although the U.S. Army had a vest and body armor at the time, the Marines defiantly paved the way in development of body armor and protection. The vest featured a web of eyelets along the bottom where items with the older style M1910 hook type wire hanger could be attached, along with a rope bridge for the rifle sling and a single pocket on the left side big enough for a notebook or pack of smokes. Many Naval personnel stationed on PBRs, swift boats, and other small crafts such as junks often wore these for protection.

Weapon is the M-79 grenade launcher, which came into service in the early 60s. Around his waist is a cotton bandoleer holding 6 M-79 40 mm rounds, each secured in a white plastic "egg-crate" holder. Due to the limited number of grenade vests available, many American Advisors like this one had to improvise, adapt, and overcome. Rounds for the weapon were usually carried in the unsatisfactory method of the cotton ammunition bandoleers. The cotton bandoleers were flimsy, ripped easily, and

Close-up shows the advisor's M-79 grenade launcher and his single bandoleer containing six 40 mm high explosive rounds. This also provides a good look at the Navy advisor wearing the Marine Corps M1955 flak vest.

wore through rapidly, and could not accommodate the grenadier with enough rounds for his basic fighting load (27 round), which meant the advisor would have to carry 5 bandoleers. Grenadiers complained that carrying several bandoleers also created uncomfortable weight imbalances and hindered easy one handed access to the rounds in their bandoleer pockets."[149] Some soldiers remedied this by utilizing empty Claymore bags or empty ammo cans onboard the junks. Since the advisors were not "humping" the extra weight of the 40 mm rounds, they could afford the luxury of bringing as many rounds of ammunition as they had available, or deemed necessary.

U.S. NAVAL ADVISOR; RIVER ASSAULT GROUP (RAG) 25 1965-66

"In the fall of 1945, the Vietnamese Army attempted to take control of My Tho, the capital city of the province in the Mekong Delta. Viet Minh forces had sabotaged the only road, Route 4, and the French vehicles were bogged down in the less than solid ground of the muddy Delta. To overcome this difficulty, the French naval inventory were brought in by small craft and successfully seized My Tho. In the two subsequent sieges of Vinh Long and Can Tho To, the waterways again proved to be the only tactical way to transport the French troops. These successes impressed the commander of French military forces in Indochina, General Philippe Leclerc. He ordered the establishment of a permanent flotilla of craft capable of transporting the Naval infantrymen where needed in the Delta. This initial force was called the Naval infantry river hotel, but as the concept was developed and more flotillas were created, the name was changed to divisions navales d' assault (naval assault divisions) and abbreviated to Dinassauts."[150]

"To strengthen the river force in the critical Mekong Delta region, on 17 March 1964 President Johnson approved use of MAP funds to create a 7th River assault group. The Secretary of Defense Robert McNamara raised the possibility of another Vietnamese RAG during his visit to Vietnam in December 1963. Initially, Admiral Felt and Capt. Drachnik opposed creation of this unit because of the low operational rates of existing RAGs. Gen. Harkins, however, eventually persuaded the naval officers that the shift of the Vietnamese Army division from the northern province to the Delta and an expanded River patrol mission demanded

Petty Officer Second Class, assigned to LCPL-1, one of a nine man crew, whose job it was to patrol the Dong Tranh river, watching for curfew violators, 1965.

Full figure shot of Petty Officer Second Class provides a good view of his M-14 rifle and Vietnamese made brown leather shoulder holster for his .45 pistol.

Another full figure view provides a good look at the advisor's cotton utilities.

Full figure shot of Petty Officer.

Full figure shot provides a good look at the advisor ready for the day's patrol. Extra magazines for his M-14 would be stowed on board the boat, so no web gear would be needed.

a net increase in riverine forces. When established in early 1965, the unit, designated the 27 River Assault Group, was composed of the 14th U.S. built River Patrol Craft (R.P.C.), six LCM-8s, a commandment, and a monitor. The LCM-8 was selected rather than the LCM-6, which was standard in early RAGs, because it could carry the new M113 armored personnel carriers used by some ARVN Army units in the Delta. The RPC, a 36-foot craft, was designed by the bureau ships to replace the French built STACN. The RPC's 15-knot speed was nearly twice as fast as the STCAN, and its cruising range of 200 miles was four times that of the Old French boat. Additional measures to increase the resources of the river force included augmentation of two existing RAGs with two Commandaments, 4 monitors, 12 LCMs, and six River patrol craft and transfer from the condensed river transport escort group to combat units of 12 STCANs. The later craft was especially suited to assault operations, since their V-shaped hulls lessened mine damage and their pointed bows, when wedged in mud banks, held well against the current for the debarking troops."[151]

This R.A.G. Force advisor wears the "Army's Olive green shade 107 utility shirt, produced under the specifications dated six November 1964, it was made of cotton sateen. It was designed with patch pockets and pointed flaps, buttoned cuffs, and the new style concave dull plastic buttons. The pocket design was similar

Close-up of Petty Officer Second Class advisor provides a good look at his unique water buffalo shoulder holster and his rank pinned to his right pocket flap. Note the subdued U.S. NAVY strip and nametape, quite unusual for 1965.

Close-up shows the locally made water-buffalo shoulder holster for his Colt .45 M1911A1 pistol.

to those used on Navy Marine Corps garments."[152] The matching trousers in Olive Drab shade 107 featured two rear pockets, each closed with a single button, and two front pockets. A black web belt was worn with a silver roller buckle for enlisted men and NCOs, and a gold one was authorized for officers. This Petty Officer, second class, wears a subdued machine made U.S. Navy strip above his left pocket and a matching name strip above his right pocket. Note, he wears a single large E-5 rank insignia on his right pocket flap instead of on his collar. Across his chest he wears a very crude Vietnamese made shoulder holster produced from water buffalo or pig skin for his M1911A1. Note also the separate compartment for a single spare magazine. Headgear is the U.S. Navy/Marine Corps fatigue cap in cotton sateen. Boots are the pre "MacNamara" boot that resemble the Army jump boots, and feature the toecap. Weapon is the M14 rifle.

U.S. NAVAL ADVISOR; LIEUTENANT, COASTAL DIVISION 33; JUNK FORCE JUNK FORCE BASE 33, VAM LANG 1966

"The junk force, HAI THUYEN in Vietnamese, had been created in 1960 as a paramilitary force manned by civilian regulars and led by the Vietnamese Navy. There initial idea had been to buy

Lieutenant, Naval Advisory Group, (NAG), Junk Force Base 33, Vam Lang, Mekong Delta, 1965.

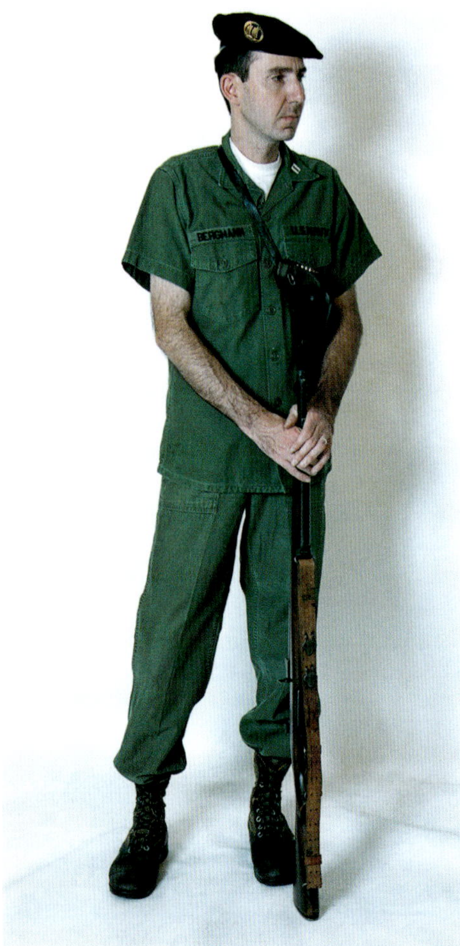

Full shot of Junk Force advisor, with emphasis on the advisor's right side.

Full figure shot of advisor provides a good view of the beret and black shoulder holster. This also provides a good look at the improved 3rd model variation jungle boots, which he is very lucky to have obtained.

Full front view shows the advisor with his M-1 rifle at the ready during operations in the Forest of Assassins, 1965.

Close-up shows the advisor's black Junk Force beret, hand sewn matching name tape, and U.S. Navy strip, along with his black shoulder holster for the S&W .38 pistol.

junks from local fisherman, but the fisherman balked, so the U.S. government contracted a private South Vietnamese shipyard to build the vessels. Because of a rush job, however, a combination of unskilled labor and green sisao wood had produced inferior vessels. By the early part of 1964 nearly one-third of them were laid up."[153] "The waterfront area of junk force bases, where the Junks are moored at night, are usually constructed by attaching planks to floating barrels. These facilities were often primitive and very vulnerable to Viet Cong attack."[154]

The American advisors and their Vietnamese counterparts usually worked from the command junks, or Chu Lucs, as they were called. "These were black hulled with a gray superstructure and no sailing masts, only a long whip antenna and a small cross mast large enough to support a flag."[155] On the front of the junk is a .30-caliber M1919 machine gun mounted to the deck.

This Lieutenant assigned to the junk force wears a set of Navy/Marine Corps cotton OG 107 uniform, and is very similar to the Army's third model OG 107 cotton sateen uniform. It features two patch pockets with pointed flaps, buttoned cuffs, and features the old style buttons. The Army's version was produced in November 1964, and was very similar, but featured the new style concave dull plastic buttons found on the newer jungle fatigues.

Close-up of Navy black leather shoulder holster, along with .38 Smith & Wesson pistol.

Close-up provides a good view of the advisor's U.S. Navy pilot's shoulder holster with added "bullet loops," and also provides a good view of his rank (Lieutenant; O-3),on his collar points.

Above his left pocket is a Vietnamese hand sewn U.S. Navy strip, while the matching strip with his name "Bergman" is worn over his right pocket in accordance with U.S. Army regulations. Worn on each collar point are the Navy's twin bars, lieutenant. In the Army or Marine Corps this would be Capt. Note, the shirt is not tucked into the matching trousers. This uniform was warm enough, without having the added discomfort of wearing the shirt tucked in.

The jungle fatigues were not a Navy issue item and therefore not in the Navy supply system. Many advisors obtained the Army's new comfortable jungle fatigues through Army supply channels, bartering, or from other Navy advisors returning home. Note that this advisor, like many others, has attempted to make the uniform cooler by converting it into a short sleeved garment. Sidearm is a Smith and Wesson .38 cal. Pistol, worn in a black leather shoulder holster. Note the addition of the extra bullet loops that have been glued into place. This advisor probably obtained the holster from a Naval Aviator, where it was common amongst pilots to add these loops to their holsters.

Headgear consists of a black two piece wool beret with the gold officer's "twin sails" junk force device. The badge is thin stamped silver metal that is chrome plated with a pin back type fastener. In the beginning, all beret badges were silver, with officer's badges being designated by a red cloth backing behind the badge. Later, about 1964, officer's badges began being made in gold. The badge was supposd to be worn on the left side of the black beret. These were usually awarded to the Americans after serving aboard the junks for a certain amount of time, or once they had been in combat while aboard the boat. The third model jungle boots with DMS are worn, which featured the new reinforced ankles and the slip-in Spike protectors, and both were introduced in 1965. Weapon is the outdated WWII M1 Garrand.

U.S. NAVAL ADVISOR, LIEUTENANT (jg)
ZONE I COASTAL, COASTAL FORCES
DANANG,
1967

"The Vietnamese had organized a Sea Force in January 1956. Shortly thereafter several American advisors were assigned to the Vietnamese Naval Base in Saigon to assist the French with on-the-job training of ship crews. Five Sea Zones were established, stretching from the 17th parallel to the Cambodian border, and one unit operated from Danang, Nha Trang, Vung Tao, Son island, and later Qui Nhon. Under their Sea Zone Commanders, Naval Posts were established at tactical points to control coastal traffic or to guard the naval territory. One post was located at the mouth of the Cua Viet River to South a 17th parallel, and the other at Khoai island off the southern tip of Vietnam. The ships

Full figure shot of Zone 1 Coastal advisor, Danang, 1967.

Full figure shot provides a good view of the Coastal Force advisor and his WWII era shotgun. Note the private purchase camera around his neck.

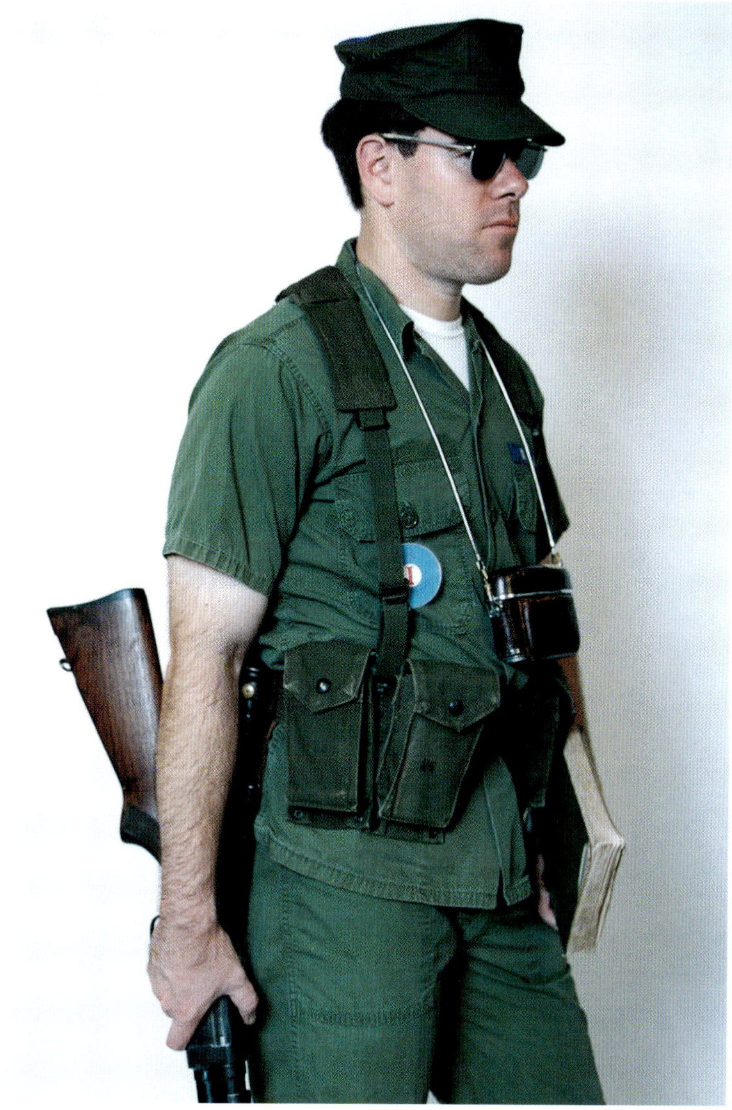

Close-up of Coastal advisor's right side provides a good view of the Coastal Group pocket hanger of the first coastal group.

and craft of the Sea Force were administratively organized into 8 Flotillas, which were in essence tiny commands. But however sound this constant patrol may have been, the Vietnamese Navy lacked the full capacity to deny communists use of the coastal waters.

"The physical difficulties of patrolling the long coastline effectively, as well as the Vietnamese limitations, were well known by American naval officers in the MAAG. In the spring of 1956 they examined a Vietnamese plan to organize 50 civilian manned, fourteen ton motorized junk forces that were to patrol in shore waters along the coast and act as an auxiliary to the Sea Force. Four years would pass before the concept was implemented. The Vietnamese Navy now began to test its sea legs during a month long goodwill cruise of two PCs (submarine Chasers) and an LS M. The squadron, under the personal command of Commander My, was well received in its ports of call—Manila, Singapore, and Bangkok."[156]

"Body armor worn by this U.S. Naval advisor is the USMC M1955 Armor Body Fragmentation Protective, or flak vest. The M1955 was an upgraded version of the M1951 vest, developed by the Corps. during the Korean War. The Marines were pioneers in the use of body armor and placed more emphasis on it than did the Army."157 Later versions of this would feature two lower pockets.

The advisor's Load Bearing Equipment is the Marine Corps M1961 web gear, or "Deuce Gear," comprising the M1961 web belt and four pouches, called M14 pouches intended for a single M14, twenty round magazine."[158] As you can see, the pouches are full of 12 gauge shotgun shells for his WWII era trench shotgun. Army M1956 suspenders are used in place of the USMC ones. Worn on his right hip is a private purchase Connetta Bowie style knife, not a K-bar. Worn on the back of his gear are two Army M1956 canteen covers with two plastic 1qt canteens and a Marine Corps Jungle First Aid Kit, which is in the center.

Uniform consists of the private purchase Genuine Trooper fatigues, which could be purchased at either the Army's PX or the Navy's BX. They were modeled after the 2nd model OG 107 cotton sateen uniform, but lightweight cotton poplin was used instead. These were intended for soldier/sailors who were not "authorized" the jungle fatigues, or who did not have access to the more comfortable uniform. The shirt featured two breast pockets, while the trouser featured two front slash pockets and two rear pockets. Above the advisor's left pocket is a locally made white on blue U.S. NAVY strip. Above his right pocket would be

Close-up with emphasis on the advisor's left side gives a good look at the advisor's Green Book of Coastal Vessels and his private purchase camera.

Close-up of the advisor shows his white on blue U.S. Navy strip still being used, along with his Marine "deuce" gear. Note his M-14 pouches are full of shotgun shells for his weapon.

Full rear view of advisor shows his complete deuce gear, complete with jungle first aid kit, two new plastic canteens with M1956 covers, and a Bowie style fighting knife, sometimes mistakenly called a "K-Bar" knife.

Close-up of Green Book of Coastal Vessels, South Vietnam.

his matching name tape, which has since been removed. The advisor has made the garment even cooler by cutting off the sleeves, typical for members of the U.S. Navy. Worn on the right pocket is a 1st Coastal Zone pocket hanger. It features a blue disk with a dark blue border, with a white circle and a Roman numeral I. There is also a dark blue anchor, which is hard to see. Other variations of this patch featured a silk woven design with the words HAI-QUAN (Navy) and DUYEN-HAI (Zone I Coastal). According to Clem Kelley's book, "There were 5 coastal zones. Each zone corresponded to the Military Region, except MR IV, which had 2 coastal zones which covered the eastern part of MR IV and CZV, the western part, which included the Nam Can & Cambodian border." Underneath, he wears a white T-shirt. Around his neck is a private purchase camera. Headgear is the Navy/Marine Corps fatigue, or utility cap. Sunglasses are aviator's, which could also be purchased, traded, or "acquired" from supply.

Close-up rear shot shows the advisor's M1961 Marine Corps pistol belt and M1956 Canteen covers with two plastic canteens. The M-14 pouches in front are full of loose 12 gauge shotgun shells for his shotgun, along with other small personal items.

Full view of advisor in battle gear, M1955 flak vest, and M1 steel pot.

A good look at the blue naval "ball cap" worn by many of the U.S. Advisors. This side view of the blue naval "ball cap" shows the difference in appearance to the standard U.S. Army "ball cap."

Full view of advisor with WWII era shotgun at the ready, helmet, and M1955 flak vest.

Full rear view of advisor geared up for battle action. Note the flak vest sits on top of the canteens.

Close-up providing a good look at the typical blue naval "ball cap."

Close-up of printed white tag on the inside of the blue ball cap.

Close-up shows the advisor in the M1955 vest and M1 steel helmet. Note the pouches are full of 12 gauge shotgun shells.

Upper torso shot provides a good look at the WWII era shotgun.

Naval Advisory Group patch, machine made.

Binoculars with case, used by the U.S. Army, Navy, and USMC.

U.S. NAVY ADVISOR; CHIEF PETTY OFFICER RIVER ASSAULT INTERDICTION DIVISION, (R.A.I.D.) 71; VAM CO DONG RIVER; 1969

Worn by this recreated R.A.I.D. Advisor is a privately purchased "Genuine Trooper" cotton poplin shirt in OG 107. These could be purchased at either the Navy BX or Army PX and worn in lieu of the heavier cotton OG 107 sateen. The shirt features two upper patch pockets, secured with a single plastic button, and in all aspects mirrored the second model sateen shirt. Above the two pockets are Vietnamese produced nametape and U.S. Navy strip. The shirt has been modified with the addition of a pen pocket on the left sleeve, and obviously been turned into a short sleeved garment, typical of naval men on the boats. Around his neck is a drive on rag, fashioned out of an OD triangular bandage. Over that he wears the Body Armor, Fragmentation Protective Vest with three quarter collar, M69, which was much the same as the M1952 vest, but the new vest did away with the shoulder loops, but added a ridged collar to protect the wearer's neck. On the vest are two M18 smoke grenades. Trousers are tiger stripe, printed on lightweight cotton poplin in an unusual pattern. They feature two large rear pockets, two large cargo pockets, two front "slash" pockets, and an odd little pocket on the left leg below the cargo pocket,

Don Slinkard (left) with unknown Naval Advisor in front of H.Q. and Don's jeep, Ben Luc, 1969. Both men wear 3d model jungle fatigue with the NAG shoulder sleeve Insignia and the black NAG beret. Note the beard on the unidentified advisor. (Don Slinkard collection)

Full figure view of R.A.I.D. 72 advisor along the banks of the Vam Co Dong River in 1969.

Full figure view of RAID 72 advisor without his M-16 rifle, on the way back to the RAID HQ at Tra Cu, 1969.

Naval advisor Don Slinkard receives the Navy Commendation Medal for outstanding service as an advisor to the Vietnamese RAG 53, Ben Luc, 1969-70. (Don Slinkard)

Naval advisor Don Slinkard returns salute to Senior Naval advisor during award ceremony at Ben Luc. Note that both the Senior advisor and the man behind the podium wear locally made black ball caps. The Senior advisor also wears a set of ERDL rip-stop fatigues. Also note the volleyball net in the corner. (Don Slinkard)

which differs from the usual cigarette/first aid pocket seen on most tiger stripe uniforms. Boots are the final model, which feature the Panama soles. Headgear is an M-1 steel pot with no cover. Weapon is the M16A1 rifle, which by this time in the war, was being issued to most South Vietnamese Forces, including the RAG Forces.

U.S NAVAL ADVISOR; PETTY OFFICER SECOND CLASS;
NAVAL ADVISOR GROUP (NAG)
R.A.G. 53; SAIGON
1969

This NAG advisor wears a mismatched set of locally produced tiger stripe fatigues. The shirt is a "watery" color, which although unseen is a sleeveless model. It features two small breast patch pockets, each secured with a Vietnamese plastic button, and lacks the side billows, which makes the pocket very small and useless. The tigerstripe pants are printed on lightweight material much like his shirt, and feature two large cargo pockets and two rear pockets, but lack the front two "slash" pockets common on American garments. Naval Advisors, like this re-constructed one, were allowed to privately purchase Vietnamese camouflage clothing like tiger stripe, ARVN ranger pattern, "duckhunter," and even the brown National police pattern for wear, without being reprimanded from higher-ups. Some even took to growing full beards, long mustaches, and "mutton-chop" side burns. The U.S. Advisors were usually left alone to do their jobs, free of harassment. Worn over his tiger stripe fatigues is a locally produced raincoat, made from the Army's standard nylon coated poncho, with hood, Olive Green shade 207, which was adopted in 1961. The raincoat is cut long, reaching his knees, and is big enough to accommodate his "deuce" gear (If he were wearing any). The raincoat features two large cargo pockets and is lined with WWII era camouflage para silk. Headgear is the ERDL camouflage boonie hat in "rip-stop" cotton poplin, which became available in 1968; this one is also privately purchased. Weapon is the M16A1.

Close-up shows the RAG 23 advisor as he would appear aboard ship with a private purchase locally made rain coat and M-16 rifle, 1969.

Close-up shows the advisor's M-16 rifle and raingear. Note the parachute silk liner and his ERDL boonie hat.

Additional photo provides a good look at the RAG advisor's mismatched set of tigerstripe fatigues. Ben Luc, 1969. (Don Slinkard)

U.S. NAVAL ADVISOR; PETTY OFFICER SECOND CLASS; NAVAL ADVISORY GROUP (NAG) R.A.G. 53; BEN LUC, 1970

Worn here by this Naval Advisor is the ERDL camouflage tropical uniform in the "short lived" poplin cotton version. The uniform is identical to the third model OD cotton poplin tropical uniform. First introduced to Vietnam by U.S. Army for use by its Pathfinder and Recon units, it would eventually be issued to most Marine line units and select Army Infantry units by 1969. Although not regularly issued to most U.S. Navy units, NAG personnel were allowed to purchase these uniforms at their own expense and wear them, although others were able to obtain them through Marine Corps supply channels, or through "trading," i.e. captured V.C. weapons, or flags. Headgear is a matching ERDL boonie hat in rip-stop cotton. These hats were favored over the black wool NAG berets while on missions. This advisor, a mem-

Naval advisor Slinkard with jeep at Ben Luc, late 69-early 1970. He wears a set of ERDL ripstop fatigues with matching ERDL Boonie hat, along with a .45 on his hip. (Don Slinkard)

Full figure view of NAG advisor to the South Vietnamese RAG 53 searching a junk for illegal items, 1969- 1970.

Full view of American advisor on boarding party detail with late 1950s mine detector.

Naval Advisor Don Slinkard on board a junk looking for contraband. Note the patch of the 53d RAG force on his sleeve. (Don Slinkard)

Advisor Don Slinkard shows off captured AK-47 rifle with bayonet. Note the tigerstripe pants and black web belt with the roller buckle. (Don Slinkard)

Full view of advisor provides a good view of the mine detector and its battery pack.

Close-up provides a good look at the back of the outdated mine detector.

ber of a boarding party, demonstrates some of the unique and outdated items that Americans had to contend with every day. During routine stops, the PBR with the American Advisor onboard would pull aside the Vietnamese craft, while one of the Vietnamese sailors would use a bow hook to secure the suspect boat. Another Vietnamese sailor would keep the occupants covered with a rifle or one of the PBR's M-60 machine guns. Members of the boarding party would be armed with pistols only, if any weapon at all, to allow them free use of their hands to search the boat, and/or operate the mine/metal detector. Rifles were not carried by members of the boarding party for fear of having them taken away and used on friendlies. Sometimes a South Vietnamese Liaison Officer or the interpreter was part of the boarding party. After securing vessels and finding nothing out of the ordinary, crews would give out items, such as candy, soap, and medicine, for a "reward." Sometimes local fisherman would catch wind of the reward system, and numerous boats would flood the search

Right: *Don Slinkard with interpreter Kim. Both men wear tigerstripe fatigues with leaf pattern boonie hats. Don wears a civilian double buckle leather watch-band, which were very popular in the 60s and 70s. Note Kim wears civilian "fruit boots." Ben Luc, 1969. (Don Slinkard)*

Don and Kim "horse around" on board ship during a break. Note the civilian boots and the radio, along with the can of Dad's root beer. (Don Slinkard)

Both Don and his friend Kim after their "Mortal" combat, and still friends. Both are enjoying a warm can of Dad's root beer and listening to the radio. Note Kim wears an ERDL pattern beret. (Don Slinkard)

Don Slinkard assigned to RAG 53, on board ship with a mismatched set of tiger stripe fatigues and an M-1 carbine. Note the salty tiger boonie hat. (Don Slinkard)

Don Slinkard with friend enjoy a smoke break. Don wears matching tigerstripe fatigues with a single dog tag on a shoelace around his neck, jungle boots, and a Montagnards brass bracelet, 1969. (Don Slinkard)

Don Slinkard sits on top of the gun turret with his M-16 in hand. Note the bandoleers of ammo worn "Poncho Villa" style and the U.S. flag in the background. (Don Slinkard)

Don Slinkard wearing locally made rain jacket and armed with his M-16 rifle on the river. Ben Luc, 1969. (Don Slinkard)

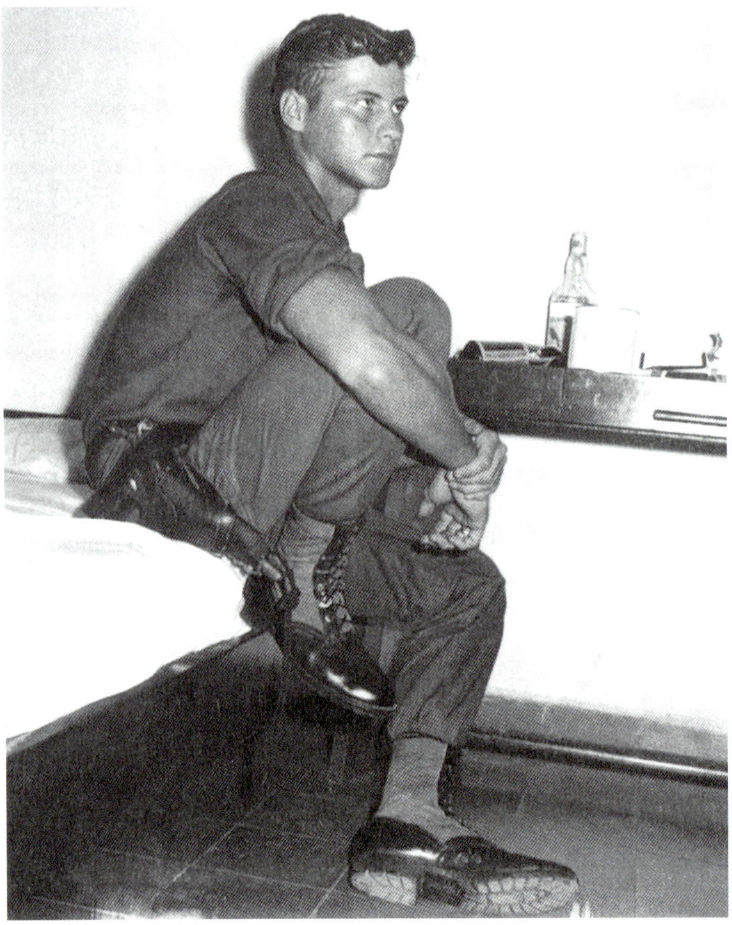

RAG 53 advisor in Naval advisor compound, Saigon, 1970. He wears a set of cotton OG 107 sateens, along with some highly spit shined jungle boots. Note the Vietnamese made cowboy holster and the numerous bottles of hard liquor behind him. (Don Slinkard)

RAG 53 Advisor wearing OG 107 cotton utilities, an ERDL beret, and a "six gun" on his hip. Note the cot behind him with the Kodak camera and fatigue cap, Saigon, 1970. (Don Slinkard)

Numerous South Vietnamese sailors wearing a wide variety of uniforms. One wears tigerstripe, while the others wear OG 107 and a mixture of South Vietnamese Naval hats.

area in hopes of "free booty," which would hinder the effectiveness of the searches. The mine/metal detector on his back is a U.S. Army Detector set, an AN/PRS-4, made by the Emerson Radio & Phonograph Corp. DA 44. On his right hip is a M1911 45. Cal pistol in the black leather holster.

U.S. NAVAL ADVISOR; PETTY OFFICER SECOND CLASS, COASTAL GROUP 23, "JUNK FORCE" SONG CAU HQ, 1970

"An American advisor to the Vietnamese Navy junk force must be many things—teacher, mechanic, Coxswain, Cook, weapons man, perimeter defense expert, first aid man, and goodwill ambassador. To be effective, he must also adapt many local customs and learn a working knowledge of the Vietnamese language.

Junk force advisors are probably the most isolated sailors in Vietnam. Some 100 advisors serve with 20 of the Vietnamese Navy coastal groups located along 1,000 miles of coastline in South Vietnam. Many of the bases are accessible only by air or water. Usually three to five advisor serve with another 25 Vietnamese sailors at each coastal group. An average of 10 junks—usually about eight Yabuta junks and two large Chu Luc command junks—are found at a coastal group. These 50 ft. wooden craft are converted fishing junks armed with 50 Cal and 30 caliber machine guns, plus a variety of small arms.

There are both officer and enlisted advisors in a coastal group. The officer assists the Vietnamese Commanding Officer in planning and executing operations, and acts as liaison when the junks operate with U.S. forces. Enlisted advisors, usually boatswains mates, gunners mates, or enginemen, worked closely with the Vietnamese senior enlisted men, giving them guidance and advice in seamanship, gunnery, and engineering. Many coastal groups have hospital corpsman to give medical attention to the remote villages near the bases. U.S. Naval advisors have been riding Vietnamese junks for nearly 15 years, and first came to assist a small Navy after the French left in 1954. During the years, the advisor's roll has changed little. The hours are long, the bases still isolated and lonely, and patrols still often consist of hours of boredom punctuated by moments of fear and terror."

Some 50,000 sampans and other small craft operate in South Vietnamese waters each day. Most are engaged in fishing and friendly trade, but not all. Some are used by the Viet Cong to infiltrate men and supplies, and it is these the junk sailors must intercept. Thousands of craft are stopped and searched each week, making it extremely difficult for the enemy to transport his war materials. Junks also provide troop lift and fire support for the Vietnamese Army and regional force troops. Advisors encourage the Vietnamese to accomplish as many of their tasks as possible

Full figure U.S. Navy advisor, Naval Advisory Group, Junk Force, Coastal Group 23, Song Cau, 1970.

A very young Lieutenant (jg) Watson, assistant advisor to the Vietnamese Coastal Group 23. He wears an OD green T-shirt and rip stop trousers that have been cut and rolled up above the knee, along with a black Naval advisory group (NAG) beret instead of the junk force beret. Song Cau, 1970. (Steve Watson collection)

Typical Vietnamese junk painted gray, along with two Vietnamese sailors. Note Vietnamese sailor's uniform consists of a pair of shorts. (Steve Watson collection)

Lieutenant (jg) Watson fills sand bags outside of coastal group 23. He wears the final pattern jungle fatigue trousers in rip stop, along with an ERDL boonie hat and a cotton bandoleer of ammo around his waist for his M-16 rifle. Son Cau, 1970. (Steve Watson collection)

Lieutenant (jg) Watson wearing a set of ERDL camouflage he obtained from an outgoing advisor. Note the ERDL uniform has been made into a short sleeve garment, and trousers had been cut off at the knees to make a pair of "Bermuda" style shorts. Headgear is a black beret with unidentified beret flash. Standing next to him is one of the junk force sailors, wearing a typical gray cotton Navy uniform, along with a ERDL camouflage boonie hat, which is modified to a short brim. Coastal Group 23 Base, Song Cau, 1970. (Steve Watson collection)

Full figure view of Coastal Group 23 advisor with emphasis on his right side provides a good look at his M-79 grenade launcher and .45 cal pistol in the black leather holster. Note the odd tiger stripe pattern cowboy hat he carries in his left hand.

Full figure junk force advisor, Coastal Group 23. Many of the U.S. advisors dressed a little more conservative while around the "brass." By this time in the war, many of the advisors assigned to the Coastal and Junk Forces wore longer than average hair, Fu Man Chu mustaches, mutton chops, and even full beards.

without assistance. In most cases, during an operation, advisors try to limit themselves to coordinating air and fire support. Camp administration and perimeter defense are totally in Vietnamese hands.

Since the junk force was incorporated into the Vietnamese Navy in July 1965, it has taken on a more military appearance. Junks are now painted in a uniform gray, although they still retain the traditional eye of the Dragon on the bow. The junk sailors believe the eye wards off evil spirits. Although the junk sailors still cling to many of the ancient ways, they quickly adapted to modern warfare. Many of the Vietnamese sailors in a coastal group are ex-fishermen, and were born and raised in the area of operations and know it well. They are a tough, hardy group. Many of them have been fighting the Viet Cong for years and have a fierce hatred of them. It is not uncommon to see a junk sailor with "Sat Cong," kill communists, tattoos across his chest, even though he knows it will mean immediate execution if he is captured by the Viet Cong.

Supplies continue to be the big problem, but most advisors agreed this will eventually be corrected. After all, said an advisor, the U.S. Navy would have a supply problem also if we more than doubled our manpower in the two years as the Vietnamese Navy has. The Vietnamese Navy has grown from 17,000 to almost a 40,000-man force in less than two years. Being far removed from the supply system, advisors often have to depend on Army helicopters, Coast Guard boats, their own ingenuity, or the old Navy art of "Cumshaw." The term "Cumshaw" originated with American sailors in China during the 1920s, and loosely means trading something you have in excess for something you need. Engine spare parts are sometimes weeks away, so advisors must become mechanics to keep their equipment operating. So advisors seem to enjoy grumbling about the time spent keeping their equipment working. The humid climate plays hell with their equipment; often over half of the advisor's time is spent repairing it. Sometimes advisors must take a few days off for a "Cumshaw" trip for spare parts to Saigon, Que Nhon, or Danang. Advisors may live in any type of housing, ranging from an Old French dispensary, to a Seabee built "Butler building." Radios, refrigerators, tape recorders, and occasional television sets in the less remote bases add a touch of home to the "hootches."

Generally, because of tropical climate and dislike for cooking, advisors eat only one meal a day, usually in the evening.

Close-up view provides a good look at the advisor's load bearing equipment and his bandoleer, which secures six rounds for his M-79. Note the locally produced pocket device with "Sat Cong," or Kill Communists written on it. Also note the subdued rank insignia on his collar.

While on patrol they share the food of the junk sailors. Occasionally, to the delight of all concerned, a new advisor will report who likes to cook and is good at it. But this is the exception rather than the rule. Power usually comes from a generator shared by the entire base. The generator is operated only a few hours at night to conserve fuel, which must often be flown in by helicopter. Many coastal groups are lucky enough to have Army helicopters bringing movies with some degree of regularity. As soon as it is dark, a screen is set up somewhere on the base, and Vietnamese sailors and their dependents enthusiastically watch the flicks. Although they do not understand what is said, it does not seem to lessen the enjoyment. Obviously, Westerns and slapstick type comedies are the most popular, along with Disney cartoons.

But, despite the isolation and inconvenience of the job, most advisors would not have it otherwise. They try the worker challenged to their ingenuity and resourcefulness. Isolation means they can accomplish their tasks in their own way. This is expressed best by one advisor who said, "Here you can definitely see the results of your efforts; it's all around you, you live with it 365

Three additional advisors from the 23rd Coastal Group at Song Cau. At left is Lieutenant Mike Lingo, wearing the final model jungle fatigues in rip stop with a camouflage cowboy hat and carrying an M-16 rifle/XM-148 grenade launcher. Also note the full beard. Next to him is GM1 Robert Odom, wearing the Navy's OD cotton sateen uniform with short sleeves and trousers that have been made into shorts. In his hand is a 12 gauge shotgun. The final advisor, at the far right, is EN 2 Robert Pearl, wearing the Navy's dungarees and an M1 helmet with the Mitchell pattern cover. Over the shoulder he carries an M-16 rifle. This photo shows the diversity of uniforms worn by the junk force advisors in 1970. (Steve Watson collection)

Close-up provides a good look at the advisor's odd pattern tigerstripe cowboy hat.

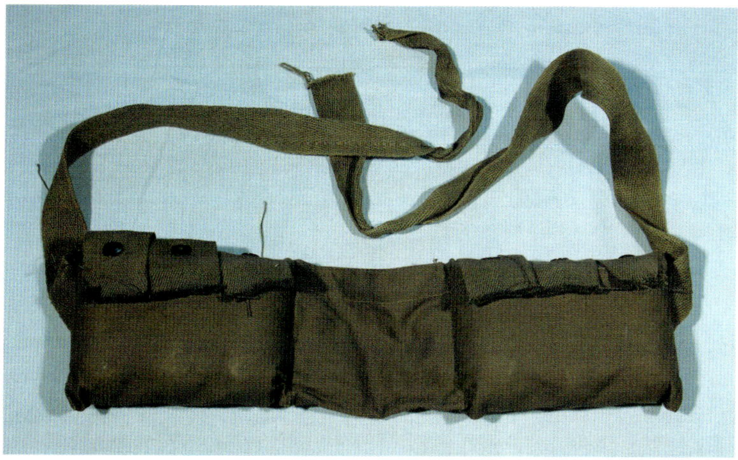

Close-up of "Junk Force" advisor's M-79 bandoleer.

Close-up of Junk Force patch (left) and 1st Coastal Group (right). Both are printed variations.

days, and your life depends on how well in the end you do your job."[159]

This junk force advisor wears a set of the 3rd model cotton poplin jungle fatigues, which have been modified to a short sleeve garment. Above his right pocket is a Vietnamese made U.S. NAVY tape, which has faded to an off brown. Worn above his right pocket is a locally produced nametape. Pin-on subdued E-5, Petty Officer Second Class rank is worn on both of his collar points. The uniform insignia worn by Navy advisors, as well as all Naval personnel, reflects their authority and shows at a glance officer and enlisted rank and enlisted ratings. Commissioned officers wear a device on their hats. Worn on his left pocket suspended from one of the buttons is a locally produced pocket hanger of the junk forces. It features an anchor, rifle, and sword that are black with white detail, on a red shield with a black border. The junk, in the center, is yellow with black detail. The words, "SAT CONG" (Kill VC) are white with black out-lines." The pocket hanger is a printed variation encased in plastic. Note, the advisor's uniform is quite conservative compared to the other junk force advisors, which means he has recently been assigned to the junk fleet, or he is close to the rear area and around the "brass" who do not look favorably on the advisors who go "native."

Don Slinkard with jaunty beret and tigerstripe uniform, inside advisor's compound, Saigon, 1970. (Don Slinkard)

LBE consists of a U.S. Army cotton H-harness connected to an M1956 pistol belt, two universal pouches, each with several M-79 40 mm grenade rounds, and a post 1967 M1956 canteen cover, which features the nylon trim with a single plastic 1qt canteen. On his right side is his Colt .45 pistol in the black leather holster. Worn across his chest are six additional M-79 rounds for his 40 mm grenade launcher. Additional rounds would be kept on board the junk in an ammo can, secured where the crew could get to them during a fire fight. In his hand is a very odd pattern tigerstripe cowboy boonie hat that actually resembles a real tiger. Third model jungle boots are worn.

U.S. NAVAL ADVISOR; PETTY OFFICER SECOND CLASS, NAVAL ADVISORY GROUP (NAG) R.A.G. 53; SAIGON 1970

This recreated Petty Officer wears the Olive Green shade 107 utility shirt that was adopted on 6 Nov 1964. The third model utility shirt was made out of cotton sateen and featured two breast "patch" pockets with pointed flaps, which gave the uniform a smarter appearance. The new Army/Navy shirt also incorporated button cuffs and the new 1963 style concave dull plastic buttons, which are still in use today on all military fatigues. The pocket design was similar to those used on other Navy and Marine Corps. Garments. The sleeves of this shirt have been modified, making it a short sleeved garment, not uncommon to U.S. Navy Advisors. Worn above his left pocket is a Vietnamese made U.S. Navy strip with a matching name tape above his right pocket. Below his Navy strip is a single "crow" E-5 Petty Officer Second-Class pin-on rank device. A pair of these were designed to be worn on

Full figure view of Naval Advisor Group member assigned to RAG 53, Saigon, 1970.

Full view of RAG 53 advisor gives a good look at the silver NCO belt buckle and rank on his pocket flap.

Full view of advisor provides a good look at the Vietnamese made "cowboy" holster for his "six gun."

Close-up showing the advisor's poplin ERDL beret.

Numerous Vietnamese sailors swab the deck. They are all wearing Vietnamese copies of the dungarees, which consist of light blue dungaree shirt, denim dungaree bell-bottom trousers, black low quarters, and the white "Dixie cups." (Don Slinkard)

Close-up provides a good look at the advisor's cowboy holster for his S&W .38 pistol, as well as the locally made name and U.S. Navy tape. This also provides a good look a the advisor's ERDL. poplin camouflage beret.

Close-up of the Vietnamese made "cowboy" holster and "six gun."

the collar points, but this advisor has chosen to wear only one on his pocket flap. His S&W 38. Cal pistol is kept in a crude pigskin "cowboy" holster which features numerous bullet loops for spare 38. rounds, just like the "Old West." Headgear is poplin ERDL camouflage beret with no rank or insignia on it. The beret was a private purchase item, and along with other headgear was often worn. Camouflage berets like this were worn with the NAG flash (usually found on the black beret) on them, as well as rank insignia, beer can D.I.s, and anything else the Advisor decided looked "Jaunty" or "salty."

Left: *Don sits on his bunk, wearing his ARVN Ranger pattern uniform and looking like the Marlboro Man of Vietnam, Saigon, 1970. (Don Slinkard)*

Full view of NAG member, Ben Luc, 1970.

Full view of Naval advisor provides a good look at his locally made holster for his .45 and NAG patch.

U.S. NAVAL ADVISOR; PETTY OFFICER SECOND CLASS, NAVAL ADVISORY GROUP (N.A.G.) RAG 53 BEN LUC, 1970

"According to Mr. Shelby Stanton in his book, US Army Uniforms of the Vietnam War, Camouflage clothing was introduced to the U.S. Army during WWII, and a post war study of the spectral characteristics of commercial dyes made it possible to formulate both visual and infrared camouflage. The Army's standard camo cotton cloth in Vietnam was of two types: printed, and water repellent, with Quarpel. This pattern was first introduced by the Army Research and Development Laboratory in 1948, and was known as the ERDL camouflage. This irregular free form, four color pattern consisted of Yellow Green, Dark Green, Brown, and Black. The ERDL was later produced in rip-stop cotton poplin in two patterns Brown Dominant and Green Dominant." On the right shoulder of his ERDL uniform is the SSI of the Naval Advisory Group, which incorporates the colors of S. Vietnam, two wreathes and two crossed anchors, both of which are white. The patch is machine embroidered. U.S. Navy and matching nametape, BAKER, are Vietnamese made and are purchased on

Interpreter Kim and unidentified Vietnamese sailor enjoy a break, Ben Luc, 1970. (Don Slinkard)

Close-up of Vietnamese made "cowboy" holster for the .45 pistol.

Close-up of Naval Advisory Group member gives a good view of his NAG patch.

the local economy. Worn around his waist is a Vietnamese made "cowboy" rig for his Government .45 cal pistol. These were produced in low-grade water buffalo, or pigskin, and could be bought at any of the local tailor shops. Across the back of the belt are numerous slots for .45 cal ammo, along with a small pouch on the holster itself for a spare clip. In his hands he carries an M-14 rifle. Footwear is the final model jungle boots with the Panama soles.

Close-up of hand sewn U.S. NAVAL FORCES VIETNAM; note the difference in shape from the previous one.

Don Slinkard takes time out of his busy day for a quick photo, Ben Luc, RAG 53, 1970. (Don Slinkard)

Full figure view of Naval advisor (NAG) to the South Vietnamese RAG 53, Ben Luc, 1969.

Full figure view with emphasis on the (NAG) advisors' right side, which provides a good view of his SSI and NAG beret flash.

Full figure view of RAG 53 advisor, Naval Advisory Group (NAG).

NAVAL ADVISORY GROUP (N.A.G.)
DA NANG
1970

Worn on the right shoulder of this re-created NAG Advisor's OD rip-stop cotton jungle fatigue is the official Naval Advisory Group, Vietnam shoulder sleeve insignia (SSI). The patch features a dark blue shield, a South Vietnamese flag, yellow wreaths, and crossed anchors. Stars and inscription are white. NAG was Head Quartered in Saigon and had Advisors serving throughout Vietnam, advising the Vietnamese Navy. A variation of this patch had white wreathes and crossed anchors, and was said to be the first variation, this worn here being the latter. Worn over the two pockets is a Vietnamese produced U.S. Navy strip and nametape, both worn with the slant of the pocket. Black Naval Advisory Group beret is worn here with the official NAG beret flash, which is a black background, yellow border, with a white "crow" on a yellow and red background, signifying the flag of South Vietnam. This flash is often mistaken for the 5th SFGA beret flash, the only difference being the addition of the white crow, and the border is yellow, not white like the Army's flash.

Close-up showing black Naval beret with the machine made NAG flash with direct embroidered white crow in the center. Note the maroon burlap liner.

Full view of advisor's right side clearly shows the NAG patch.

U.S. NAVAL ADVISOR; LIEUTENANT, ADVISORY TEAM 159, COASTAL GROUP 43; JUNK FLEET; FOURTH COASTAL ZONE; HA TIEN, 1970-1971

Ha Tien was a small arm of land, with a shallow harbor based at the mouth of the Giang Thanh River. The American Advisors worked out of a 50-year Old French plantation house that had a steep-pitched tin roof with a view overlooking the bay. The advisors from both Coastal Group 43 and 44 are headquartered here. Along the shore are several thatched houses and a few two-story buildings. The Advisors at Ha Tien lived in the Stone Age; the building offered no running water or flush toilets, and they usually slept on the porch on a cot with mosquito netting to take advantage of the sea breeze. Their mission: to join with the American Forces still in country, patrolling the Giang Thanh river and Vinh Te canal, which defines the Vietnam-Cambodian border. Basically stopping the VC traffic from crossing into Vietnam. The VC used the river to re-supply his Main Force Battalions in the U-Minh forest, the Coastal Group's job was to form an inland barrier to intercept any supplies the VC tried to cross along the lower 30 miles of the border. The junks patrol the river during

Close-up of locally made NAG beret flash and pin-on E-3 rank, which was added to the beret after the Vet's tour was over. Note that this photo was taken at a reunion. Also note the owner's name inside the beret, Donald E. Slinkard, and his serial number.

Full figure of "Junk Force" advisor, Coastal Group 43, Hon Tre Island, 1971.

Full figure of advisor with the WWII era Thompson sub-machine gun.

Full figure of the "Junk Force" advisor. Note the gold twin sail junk force emblem on his black junk force beret. This angle also provides a good look at his outdated .45 cal Tommy gun.

Full figure of advisor removing extra Tommy gun magazines from his pocket, which is the only place he has to secure them.

hours of daylight and then set up ambushes along the river and the Vinh Te canal at night."[160]

The Vietnamese "Junkys" wore black PJs much like the civilian fishermen in Ha Tien, standard M69 flak vests, U.S. M-1 helmets, and their rifles, which by this time were M-16s. The American advisor, seen here, a LT would usually wear standard OG 107 jungle fatigues, preferably short sleeves, with U.S. jungle boots, an M-69 flak vest, and a helmet and an M-16 or M-79 grenade launcher. Some advisors chose side arms such as the .38 cal Smith & Wesson worn in a shoulder holster or on the pistol belt in either a U.S. made holster, or a locally produced one. Headgear was usually anything you liked, such as the advisor here, wears a "cowboy" style boonie hat in the early ARVN Ranger pattern. Some U.S. Advisors wore these style cowboy hats, while others wore the traditional black Junk Force beret with either a gold Junk Force device (for Officers), or the silver device for enlisted men. The U.S. Advisor was also responsible for the AN/PVS Starlight scope that was kept in its aluminum case until needed, along with the PRC-25 radio, which was used for air-support (Seawolves) and to check in with the Naval Operations

Close-up of junk force beret with the officer's gold twin sail junk force device; also note the blue liner.

Close-up provides a good look at the Junk Force advisor's black pajama shirt. Note the high collar, the buttoned cuffs, shoulder loops, and the pocket flaps are styled after U.S. second model OG 107 cotton fatigues.

Close-up of the gold officer's "twin sail" Junk Force beret badge.

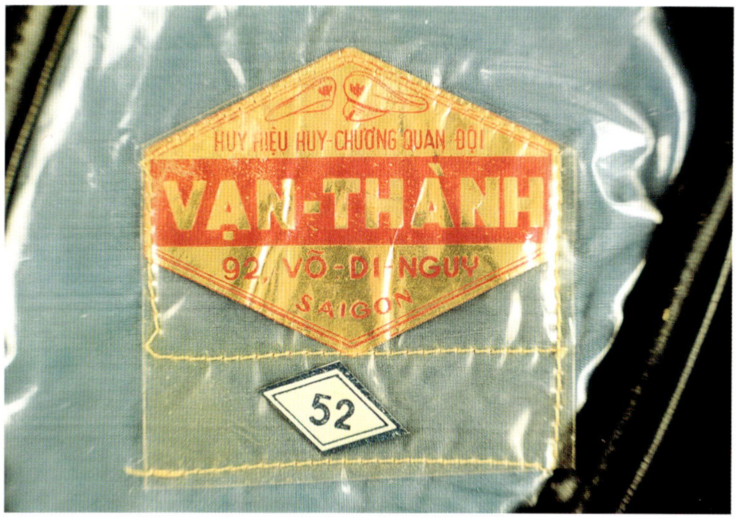

Close-up of the U.S. advisor's "Junk Force" beret's blue liner and Saigon manufacturer's tag, common in most South Vietnamese berets, because this was the largest hat maker in the South.

Center, located aboard the USS Garret County, anchored two miles offshore."[161] The patch for Advisory Team 159-Fourth Coastal Zone, was a purple dragon wrapped around a sea anchor over a background that looked like the South Vietnamese flag. When the Team moved the Coastal group to the small island of Hon Tre, no one was around to question their attire, so the U.S. Advisors went "native" and began wearing the black PJs when at the base, and on patrol as well, although the standard issue OG 107s were still worn, also. The flak jacket was always taken on operations, but usually only worn if they knew they would be in contact, most of the time it was used as a pillow or cushion to sit on. The

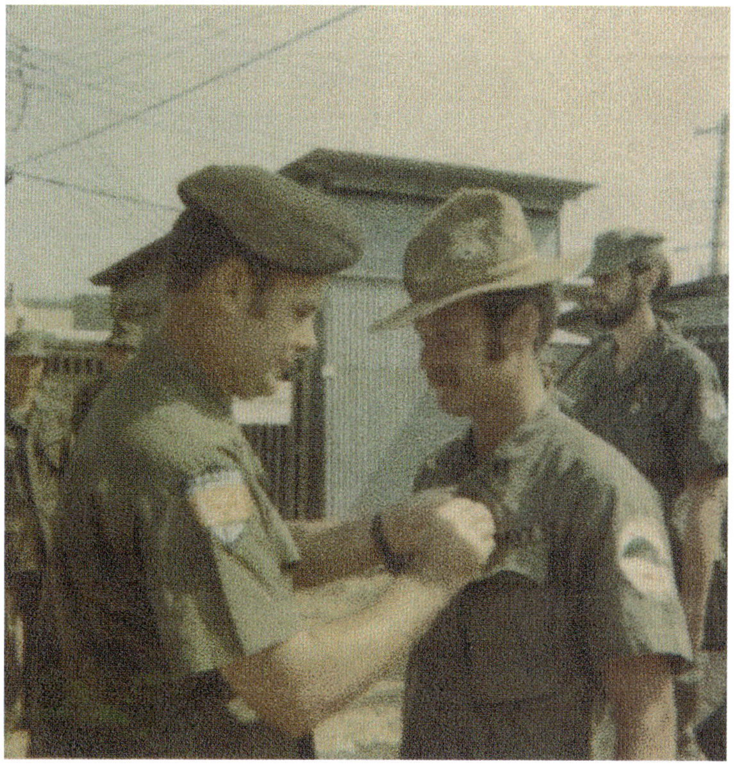

Senior Naval Advisor to IV Corp, Commander Shine (left with black "Junk Force" Beret and NAG patch) pins award to Lt. Sam Lee Pfiester, April 1971. Note his crude cowboy style hat and the circular "Junk Force" patch on his left sleeve, which reads "Hon Tre Junkies." Also note the Naval Advisor in the rear with a full beard. (Sam L. Pfiester Collection.)

Sam Lee Pfiester (author of The Perfect War*) being read his award by Commander Shine, April 1971. Note the advisor behind Sam wears the Navy's cotton utilities. (Sam L. Pfiester Collection)*

wearing of the black PJs seemed to correspond with the amount of time "in-country." The members of Coastal Group 43 designed their own patch, as well. It was a circular patch, and featured a green turtle shaped island (which is what Hon Tre looked like), on a white background with the words "Hon Tre Junkies," along the top in red letters. This was also silk "bevo" Woven and was worn on the left pocket.

U.S. NAVAL ADVISOR; LIEUTENANT, COASTAL GROUP 44, "JUNK FORCE" IV CORPS. HA TIEN, 1970

This junk force advisor wears a slightly mismatched set of the infamous tiger stripe uniform. The shirt is printed on the lightweight poplin material, with two breast pockets, each with a billow on the inside. The pockets are closed with a single concealed button. The tiger stripe pants worn are a strange pattern that feature two front "slash" pockets, two rear pockets, and two large cargo pockets. What makes them strange besides the color and stripe configuration is the third leg pocket, which is secured with two mini buttons and is slightly bigger than the dressing/cigarette pocket found on certain tiger stripe garments. The tiger pants have

Full figure view of the Naval advisory team member assigned to the junk force, Coastal Group 44, An Thoi, South Vietnam, 1971.

Full figure view provides a good look at the Junk Force advisor's Aussi hat.

Full figure view provides a good look at the advisor's Aussi Bush hat and .38 in a cowboy style holster.

Full figure of advisor in duty uniform, away from prying eyes of the brass.

Full figure advisor preparing for possible action aboard his junk. Note the addition of the M1 steel helmet and M69 flak vest.

nearly the same colors as the shirt; an orchard green, gold/brown and black, but the pants are clearly a larger pattern.

Around his waist is a crude Vietnamese made "Cowboy" holster and belt for his 38. Cal pistol. Headgear is an Australian felt campaign hat, or boonie hat, along with two Aussi Army regimental pins for the tank corps, and the seal of the army. Many of the Americans chose wild and colorful uniforms and headgear, and this one is no exception.

While working onboard the boats, the Aussi hat would be traded for his M-1 steel pot and the comfortable M69 flak vest, along with the addition of an M-16 rifle or an M-79 grenade launcher. Worn around his neck is the OD green bandage used as a "drive on" rag.

Advisor with the addition of his M1 helmet and ready for possible hostile actions.

Full figure of the American advisor aboard command junk along the coast. Most of the time, the flak vest worn here was used as a cushion for sitting on throughout the patrol's duration.

U.S. NAVAL ADVISOR; LIEUTENANT, COASTAL GROUP 43, "JUNK FORCE" HON TRE ISLAND; 9-FOXTROT PATROL AREA 1971

"The junk fleet of Coastal Group 43 consisted of all fairly new Yabuta class junks, painted gray, and featured a stand up cabin, along with two An Xuyen class junks, painted dark forest green, and one Kieng Giang class, which was said to resemble an old pirate ship. A set of colorful eyes were painted on the front of each to symbolize 'The eyes of the Dragon' and were supposed to keep evil spirits away. Each boat was usually armed with one .50 caliber machine gun mounted on the foredeck and a 30 cal. mounted aft. Some even had mortar racks amidships. Each of the Coastal Groups were supposed to contain at least five boats operating at any one time, with a crew of eight to ten Vietnamese sailors.[162] The U.S. had been patrolling the area for two years with their Swift boats, but turned the area over to the Vietnamese in early 1970 as part of Pres. Nixon's Vietnamization program.

While stationed on the tiny island of Hon Tre, just 20 kilometers from the coast of the "three sisters," advisors such as this recreated one and their junk force crew dressed in the black paja-

Close-up of advisor logo, which came from Mr. Pfiester's plaque. Their patch was the same design, except the background was white (instead of blue sky and water). This patch was worn by the Advisors of Coastal Group 43 on their left sleeve, Hon Tre Island, 1971. The patch read "Hon Tre" on top and "Junkies" at the bottom. (Sam L. Pfiester Collection)

Full figure of junk force advisor, Coastal Group 43. Foxtrot Patrol Area, Hon Tre, 1971.

Full figure of Coastal Group Advisor 43 with AN/PRC-77 radio and M-16 rifle.

mas, much like the fishermen in the area. This set of pajamas differs slightly than the regular PJs. The uniform features a high collar, two upper chest pockets, each secured with a single button, along with buttoned cuffs. The shirt also features shoulder loops. The matching black PJ trousers feature an elastic waistline and two front slash pockets, in which this advisor has chosen to hold three additional 30 round magazines for his 45. Cal Tommy gun. The set was probably designed for the Regional Force/Popular Force (RF/PF). Once again, the black junk force beret with the Gold officer Junk Force badge is worn, completing his uniform.

Close-up of Naval advisors "salty" camouflage cowboy hat.

Full figure of advisor Lieutenant to "junkies." Besides the AN/PRC 77 radio and communications, the Naval advisor was also responsible for the AN/PVS-2 night scope.

Full figure of Naval Advisor Group (NAG) member assigned to a unit along the Cambodian border, 1972

U.S. NAVAL ADVISOR, PETTY OFFICER FIRST CLASS NAVAL OPERATING BASE; BUTLER BUILDING, TAN CHAU; "OPERATION: TRAN HUNG DAO 18" 1971-72

"After a request from the Cambodian Government, Operation Tran Hung Dao 18 was initiated for the Mekong Delta Tactical Zone (MDTZ). The primary objective of this operation was organizing and coordinating ground and air support for convoys, which carried military cargo and petroleum products from Tan Chau to Phnom Penh. The vital river route passed through more than 100 kilometers of dense jungle, high riverbanks, and several narrow gorges, affording the enemy ideal ambush positions at almost any point. On 18 January 1972, Major Donald R. Carter relieved Navy Commander Arthur St. Clair Wright as senior advisor. Tran Hung Dao 18 (THD-18) area coordination and commander task group 116.15. Major Gardener, after serving nine months in the Chau Doc Province, was embarking upon an assignment that had political and cultural implications unavoidable even at the advisory level. Traditionally, Vietnamese and Cambodians neither liked nor respected one another. It was Major gardener's task to mini-

Close-up showing the black Naval beret with hand sewn NAG flash with the white crow, which looks very similar to the 5th Group flash. The NAG flash has a yellow border, while the 5th SFGA has a white border. Photos show that some Naval advisors mistakenly purchased the Army's flash and wore it instead of the Navy's NAG flash throughout their tour.

Full view showing Naval advisor with black Naval Advisor Group (NAG) beret.

Full figure view provides a good look at the advisor's left side.

Full figure shows a Cambodian flag in the advisor right hand while he adjusts his 45 in U.S. manufacture black leather shoulder holster with his right.

mize this historic antagonism through a continuing liaison and coordination effort."[163]

"As he recalled, 'only the good working relationship, in spite of ethnic differences, between the F.A.N.K. (Cambodian Army), the 4th ARVN Ranger Group, and THD-18 made this possible.' In addition to the Vietnamese Navy commitment of ships, tankers, tugs, and barges to haul fuel and other supplies to Phnom Penh, there was a constant requirement to support the ARVN ground forces that provided river bank security. Originally, the VNMC had been tasked with the duties, but, ultimately the Vietnamese 4th Ranger group of the 44th special tactical zone was assigned this mission in cooperation with the Cambodian 4th Infantry Brigade, and the ARVN Rangers were successful in preventing the enemy from closing the river supply route. Had this route failed, Cambodia would have fallen in 1971."[164]

"Major Gardener's VNN counterpart, Commodore Nghiem Van Phu, a graduate of the U.S. Army's Command and Staff College, had run convoys on the Red River in North Vietnam prior to 1954, before Gardener was old enough to wear a uniform. He was a naval professional, and he expected no less from the officers and men under him. Under the commodore's tutelage, Gardener quickly became accustomed to 'brown water navy techniques.' Major Gardener learned the customs of the Delta people, visited their families, and made every effort to improve sanitary and living conditions. The people-to-people effort took on a new meaning one day when on a sampan he helped to deliver a child. Every day seemed to present new challenges for him."[165]

"February and March 1972, allied forces uncovered a cachet of arms and supplies near outlying bases along the Cambodian border which indicated Communist infiltrators had prepared for more than guerrilla type harassment actions. Air strikes by Americans, Vietnamese, and Cambodian forces provided some protection, but the Communists, however, repeatedly made their presence known. In early 1972, there were more than 60 ambushes against shipping to Phnom Penh. During this time, the enemy sank or damaged other vessels with rockets. Enemy anti-aircraft fire also brought down three helicopters providing escort air cover. Had it not been for the professionalism of the Marine advisors and the cohesive working relationship amongst the different branches, far more may have been lost."[166]

The ERDL camo tropical uniform is shown here in its 2nd model cotton poplin/rip-stop version. The uniform was styled identically to the third pattern tropical combat uniform and included the same design features. "Camouflage clothing was introduced

Advisor folds the Cambodian flag while aboard the ship.

Full upper torso provides a good look at NAG flash on his black Vietnamese made beret.

Upper torso view with emphasis on his right side provides a good look at the advisor's black Naval beret and collar rank.

to the army during WWII, and post war studies of the spectral characteristics of commercial dyes made it possible to formulate both visual and infrared camouflage. The ERDL was technically restricted to the Army and distributed first to pathfinders and reconnaissance combat units during 1968, and later, select Marine Corps Line companies."[167] This uniform was not in the U.S. Navy supply system, and advisors would have to purchase this on the local economy, or obtain it from other Naval advisors returning to the United States, which was usually the case. Also note, advisors did not know this as the ERDL pattern; it was referred to as the Army's camouflage, or Lime Green camouflage. The advisor wears his pin-on rank opposite of the norm., with the crows facing outward. This was done in wartime only, or in a combat zone. In the states, these would be reversed. Worn above his left pocket is the Naval Submarine Badge. To obtain this qualification badge, one would have a PQS check list of 150-200 different tasks to perform, including stand watch, then the applicant would go in front of a board of 15 badge holders and be asked several questions. The badge was silver for enlisted and NCOs and gold for officers. It is referred to as the "Dolphins."

The black NAG beret is worn with the Naval Advisory Group Beret flash, which is very similar to the Army 5th SFGA beret flashes. The difference is that the Navy's flash features a yellow border and the Army features a white border. There are instances, documented by photos and personal accounts of NAG Advisors obtaining the Army flash and wearing it with the black Naval Advisory Group Beret.

"The largest Marine advisory effort was with the South Vietnamese Marine Corps. Beginning with one U.S. Marine advisor in early 1955, the U.S. Marine Advisor Unit (MAU) to the South Vietnamese Marine Corps had by January 1968 grown to an unauthorized strength of 33 personnel, consisting of 27 Marine officers, a Navy Doctor, four enlisted Marines, and a Navy corpsman. Commanded by Col. Richard L. Michael, Jr., who held the title of Senior Marine Advisor, the MAU was part of the Naval Advisory Group (NAG) in the U.S. MACV advisory organization. In Saigon, Michael maintained a small headquarters, which consisted of the assistant senior Marine advisor, Lieutenant Colonel Glenn W. Rodney, and small administrative staff. The rest served in the field with the deployed units of the Vietnamese Marine Corps."[168]

"Outside of the supply, ordinance, artillery, and amphibious specialist officers, the remaining 16 U.S. Marine advisors served with the two VNMC task force Headquarters or the six infantry battalions. Each task force was allotted a U.S. Marine Major and Captain as an advisor and asst. advisor. A U.S. Marine Captain

Upper torso shot provides a good look at the advisor's .45 in the shoulder holster, E-5 rank pinned to his collar points, and the Submarine Insignia, silver for enlisted. This is known as the "Dolphins."

Upper torso also provides a good look at the Submarine Insignia (Enlisted).

and 1st Lt. were assigned to each of the battalions as the advisor and assistant advisor to the commander. For the individual marine infantry individual advisor, it meant a continued 'nomadic lonely' life. As one marine officer wrote, it was not 'unusual' for a 'marine advisor to report abroad, undergo in-processing of two or three days; and joined a deployed unit not to return to the advisory unit for months at a time.' A senior advisor to one of the Vietnamese battalions, Capt. Jerry I. Simpson, commented that while serving with the Vietnamese he subsisted 'on the same rations' as the South Vietnamese Marines and would not see any Americans, including his asst. advisor, 'for several days at a time.'[169]

Admiral McCain's flight jacket. Note the NAG patch.

Close-up of advisor patch NAG on left side and name patch, along with Sub Warfare Badge.

Close-up of U.S. NAVAL FORCES VIETNAM, SSI. One of many patches worn by U.S. naval advisors.

Close-up of River Patrol Group 59 patch. It is a direct copy of a U.S. RPG patch, except the words "CO VAN MY" have been added, which means "U.S. Advisor," and "Hai Quan," the place. Also note the SVN flag and date.

Don Slinkard wearing tigerstripe and an M-1 helmet with a tiger stripe cover he made for it from another shirt. He is posing with his M-1 carbine, 1970. (Don Slinkard)

Don Slinkard ready for action, fully poseable Navy action figure! (Don Slinkard)

Interpreter Kim wearing a matching set of tiger stripe, slices open a coconut for the boat crew. (Don Slinkard)

Advisor Slinkard pretends to be busy talking on the radio to keep the brass off his back, Ben Luc, 1969-70. (Don Slinkard)

Unidentified Vietnamese sailor wearing tigers and a conical hat enjoys a smoke and a break during a patrol, 1970. (Don Slinkard)

6

U.S.M.C. Advisors to the SVMC 1964-1972

"As could be expected, the South Vietnamese Marine Corps attempted to pattern itself after the U.S. Marine Corps model. It consisted of a Lt. General Commandant and a small central headquarters in Saigon, two combat task forces, Task Force Alpha, and Task Force Bravo, 6 infantry battalions, an artillery battalion, an Amphibious Support Battalion, and a Training Center. Most of the Vietnamese Marine field officers and many of the company grade officers had attended at least the U.S. Marine Corps Basic School at Quantico, Virginia. A few of the more senior officers also graduated from the more advanced U.S. Amphibious Warfare School at the U.S. Marine base. By January 1968, the Vietnamese Marines numbered over 7,300 men and prided itself like its sister service in the United States on its elan and its reputation as one of the country's elite fighting forces."[170]

"Despite the similarities between the two Marine Corps, there were important differences. While its officers and some of its

Two American Marines assigned to the (MAU) Marine Advisory Unit help Vietnamese Marines survey the Delta in late 1964-early 1965. Note the Marine advisor in the back wears a set of the Marine cotton OG 107 fatigues with no rank on the collar, although it is believed he is a Lieutenant. Both men carry M1 carbines. The man in the front wears E-6 rank on his first model exposed buttons shirt, along with first model trousers.

Knealing Vietnamese Marine wearing OD fatigues with the green Vietnamese Marine Corps beret and final model beret device, which appears to be wire bullion. Note his OD uniform is void of all patches. Also note the locally made holster for what appears to be a U.S. .45 pistol on a U.S. vertical weave pistol belt.

15 early Vietnamese Marines wear a variety of OD and tigerstripe uniforms and carry a variety of weapons. Note, most of the Marines have M1 Garrand rifles.

enlisted men had received amphibious warfare training, the South Vietnamese Marine Corps actually participated in very few amphibious operations. Having its origins in the Vietnamese commando and Riverine Companies under the French, the Vietnamese Marine Corps at first operated much in the French tradition after its establishment in 1954. In fact, until May 1955, a French Officer remained in command of the Vietnamese Marines. With a growing American influence, the Vietnamese Marine's organization tended to reflect the U.S. Marine Corps with a growing emphasis upon the amphibious mission. Still, from the very beginning of their existence, the Vietnamese Marines were committed to campaigns against the Viet Cong. While still continuing riverine operations, especially in the Mekong Delta and in the Rung Sat Sector, South of Saigon, there was little call for assaults across a defended Beach."[171]

Full figure of U.S. Marine advisor to the Vietnamese Marine Brigade, 1964-65.

Full figure shot with emphasis on the Marine advisor's front provides a good look at his tigerstripe uniform.

Full figure shot with emphasis on the advisor's right side shows the advisor's gear consists mostly of U.S. Army M1956 items.

U.S.M.C. Advisor, 4th SVMC Battalion, Vung Tau 1964

"The basic advantage that the Vietnamese Marine Corps. offered was their national character. Recruited from the nation at large, rather than any one region as most Vietnamese Army Divisions were, they could be deployed anywhere in Vietnam when the situation demanded. Together with other specialist units, such as the ARVN Rangers and the Airborne, the Vietnamese Marines formed the National General Reserve. Operating directly under the South Vietnamese Joint General Staff (JGS), these units became in effect fire brigades to rush to the most urgent hot spots and put out the flames. In one sense, the most important quality of the Vietnamese Marines was their demonstrated loyalty over time to the central government and the Joint General Staff (JGS)."[172]

This early Marine Corps advisor to the 4th Battalion, South Vietnamese Marine Corps, operating out of Vung Tau, wears an early set of tiger stripe. Note the large size of the black stripes, sometimes referred to as the John Wayne pattern. The matching trousers worn here featured a zipper fly, two rear pockets, two front slash pockets, two large cargo pockets, and a small cigarette style pocket on the left leg, intended for the first aid dressing. Typically, both the U.S. advisors and Vietnamese Marines would wear the brigade shoulder sleeve insignia on the left sleeve of

Upper torso shot shows the advisor consulting his map.

The Marine advisor takes a minute to check his map of the AO.

Upper torso shot provides a good look at the advisor's tigerstripe uniform. Note the O-3 rank on his collar.

Additional close-up shows the advisor's upper torso and his "deuce gear," which is mostly M1956 gear on a Marine M1961 pistol belt.

Vietnamese Marine wearing the large stripe pattern tigerstripe uniform with Vietnamese Navy rank pinned to the front of his shirt. Circa 1966.

their uniform. This consisted of a black shield with the Marine Corps insignia (EGA), Eagle, Globe, and Anchor with yellow with black detail, and a red star with the map of Vietnam in yellow. These came in three different variations—handsewn, printed, or silk woven. An additional name strip or the U.S. Marines strip would sometimes be worn over the left pocket, also. Worn on both collar points is the Marine's rank, which is Captain. Affixed to his shirt just below his white T-shirt is a U.S. "skill craft" ballpoint pen.

Web gear consists of the Marine M1961 pistol belt, two M1956 canteen covers with two aluminum canteens, and two of the U.S. Army's universal pouches. Supporting his pistol belt is the Army's cotton H harness with an additional M1956 compass/first aid pouch. Headgear is the standard issue M1 steel helmet with an early cotton leaf pattern camouflage cover known as the Mitchell pattern. In his hand he holds a plastic coated map of his AO. Weapon is an early World War II era M1 carbine that lacks the bayonet lug, which would be added during the Korean War. Boots are the second model (DMS) jungle boots lacking the ankle reinforcements.

U.S.M.C. Advisor,
5th VNMC Battalion,
Bien Hoa,
1964

"Relying on U.S. Marine Corps advisors from the start, the VNMC, unsurprisingly, reflected this influence in its recruiting, organization, and training. Like its American counterpart, the VNMC recruited volunteers and did not draft. Its recruiting program stressed patriotism and challenged 'young men to prove themselves equal to rigorous, disciplined life.' This proved to be as effective in Vietnam as it was in the United States. Thirteen enlisted recruiting teams were located throughout the country. By mid-1971, 610 men were being enlisted monthly, which was enough to replace normal attrition and keep up with authorized strength increases. Officers were appointed from varied sources: the National Military Academy, the two-year infantry school for reserve officers, and the 12-week officer course for meritorious NCOs."[173]

This Marine advisor to the 5th VNMC Battalion, operating out of Bien Hoa, where the battalion had just moved, wears a set

Full figure of USMC advisor to the Vietnamese Marines, Delta region, South Vietnam, 1964-1965.

Full figure view of advisor with emphasis on his right side.

of the U.S. Army's new Tropical Combat Uniform, better known as the "jungle fatigues." The tropical coat and matching trouser were lightweight, cotton poplin weighing nearly half what the heavier cotton M51 uniforms the Marines were issued. The coat was a loose fitting item, designed to be worn outside the trousers and featured side take up tabs, shoulder loops, and four large pockets each secured with two "exposed buttons." This was strictly an Army item, being developed for the Army by the Army Engineer Labs. The first model, worn here (1963 pattern), featured the exposed buttons; by August 1964, these would be covered on the pockets only. Note, on the upper left pocket of the advisor's shirt is an early EGA (Eagle, Globe and Anchor), along with U.S.M.C. stenciled on the pocket. Pinned to each collar is the Marine's rank, E-6. This Marine advisor was probably lucky enough to receive

Close-up showing the early Vietnamese Marine green beret, which is a two piece beret, and the bullion crossed anchors and wreathes beret badge. This was the Naval Infantry device, and is very similar to the Vietnamese Navy Device. (Mike Tucker collection)

Close-up of early Vietnamese Marine green beret with the early cross anchor beret badge. (Mike Tucker collection)

Full figure view of advisor with emphasis on his front. Note the difference in the colors of his 1st model shirt and trousers; this is probably due to the trousers being worn more often than the shirt.

Close-up shows the early Eagle Globe and Anchor (EGA) stencil and his Marine Corps stencil, along with his World War II tan shoulder holster.

these through U.S. Army supply channels, or through some hefty bargaining with the supply Sgt., say, maybe a captured VC rifle or pistol, or maybe even a flag. Underneath his "jungle fatigues," the Marine wears the standard white issue T-shirt.

Web gear consists of an Army M1944 OD green pistol belt, supported by the U.S. Army H-harness, with a single compass/first aid dressing pouch connected to his right side. On his pistol belt he wears two U.S. Army Universal pouches (sometimes mistakenly called M-14 pouches), each with several banana magazines for his M-1 carbine. On each hip he wears the newer Army style M1956 canteen covers with two brand new plastic 1qt canteens, which just became part of the Army's inventory in 1963. On the rear of his pistol belt, unseen, is the Marine Corps. Jungle first aid kit, which contained a bottle of bug juice, several adhesive strips, small dressing, and water purification tablets. Worn across his chest is a WWII russet brown shoulder holster for his M1911A1 .45 cal pistol.

Rear view of "deuce gear" showing two M1956 canteen covers with two of the brand-new plastic canteens, along with one M14 pouch with a jungle first aid kit attached to it. Inside the pouch is the advisor's cleaning kit.

Close-up showing the advisor's "deuce gear," which is mostly U.S. Army M1956 gear. Note the pencil in his right pocket and his collar rank on his right collar.

Close-up of early printed B Brigade, which features the hero King Nguyen Hue, who conquered the north in 1789. The patch features the king on horseback with his followers behind him wearing the conical hats and holding pikes.

Close-up of locally made tigerstripe Marine Corps cover. (Mike Tucker collection)

Additional close-up of Vietnamese made Marine Corps cover with a U.S. made O-6 Colonel rank pinned to the front. (Mike Tucker collection)

Photo of a South Vietnamese Marine (Officer) wearing a set of tigerstripe with a camouflage M1 helmet and a brown leather holster for his .45. Note the yellow fourragere over his left shoulder.

Photo of U.S. Marine advisor to the South Vietnamese Marines in 1964. Both men wear the Vietnamese tigerstripe uniforms. Note the U.S. advisor wears a locally made name strip above the left pocket, along with an early Marine Brigade SSI, which is black instead of green. Note also he wears no rank on his locally made tigerstripe cover.

This advisor has been doubly lucky, and has obtained a set of the Army's Jungle boots, in the second pattern, which lacks the ankle support. This, too, was an Army item; most Marines would not see these until 1966-1967, and they could only be worn while in the Republic of Vietnam. Back in the states, these boots and Jungle fatigues were forbidden! Headgear consists of cotton Marine fatigue cap, with heavy starch, which seems to be holding up under the hot tropical sun. In his hands he carries his M-1 carbine with the longer banana magazine.

USMC ADVISOR, ADVISORY TEAM 43 RUNG SAT SPECIAL ZONE, 1967-1968

"Marine Advisory Team 43 was a tightly knit group made up of 3 officers and seven enlisted. These men were part of the Naval Advisory Group, Vietnam, and operated independently of the other Marine Advisory units who worked exclusively with Vietnamese Marines. Team 43 lived in and worked in a small, one story three room 'hootch' in the Vietnamese Regional Forces compound outside the Navy base at Nha Be, located 10 miles south of Saigon, just on the western border of the Rung Sat, the Forest of Assassins!

These Marines had a two fold propose: to provide specialized advice to LTC An in the fields of infantry, intelligence, and political warfare; and to provide assistance in these fields to the senior U.S. Advisor in the Rung Sat, Commander Charles N. Straney. The teams worked primarily with the 13 'Regional Force' companies, part of the Vietnamese home guard, along with providing coordination with visiting U.S., Thai, and other Vietnamese forces operating in the area. The Xuan Kich, the Rung Sat

U.S. Marine advisor Team 43 Rung Sat Special Zone, 1967.

Team 43 Marine advisor wearing a mismatch of tigerstripe and poplin ERDL items. Note his M-1 carbine has no sling.

Rear view of advisor provides a good look at his PRC-25 radio on the carrying harness.

Commandos, are the product of Marine Advisory Team 43 training. The Elite Reaction Force, the only one of its kind, has become the most efficient fighting force in the area in just 10 months. Team 43 is headed by Capt. David Brown, who serves as the unit's Political Warfare Advisor, along with the team's medical Advisor Navy Hospital Corpsman, who works closely with the Political Warfare section. Their job: to advise the Vietnamese commander on psychological warfare, civic action, and social welfare, along with providing support and advice to Regional Forces Cultural teams.

Three to four times a month the team moves out, alternately touring the upper and lower sections of the Rung Sat, visiting all the villages on the circuit. On these tours, the teams distribute wheat, oats, salad oil, and other commodities, along with sitting long enough to sing a few songs with the villagers. Captain Brown spent 5 weeks of language training at the U.S. Army's J.F.K. Special Warfare Center, Ft. Bragg, NC, along with a ten-week course in psychological operations training."[174]

This MAT 43 NCO wears a mismatched set of camouflage, a dark tiger stripe blouse and poplin ERDL pattern trousers. The

Close-up of advisor's poplin ERDL boonie hat with a short brim.

Close-up of Vietnamese 4th Marine Battalion. KIHN-NGU = KILLER FISH. Silk woven.

shirt is worn with the collar up for protection from the sun. "The AN/PRC-25 was a short-range man pack portable, frequency modulated (FM) receiver/transmitter, weighing just over 24 pounds with the dry cell battery. Each battery had an average life of about 20 hours. Maximum range for the radio was from 5KM with the short tape antenna and up to 8 m with the long seven section "fish pole" antenna. The radio was part valve/part transistor and had 920 channels, in which two were pre-set."[175] Note the plastic bag over the hand mike.

U.S.M.C. ADVISOR, TASK FORCE ALPHA, SAIGON, 1968

This reconstruction of a Marine advisor assigned to Task Force Alpha in 1968, wears a set of tiger stripe favored by both U.S. Marines and the Vietnamese Marines alike. The photos show that Marine Corps Advisors attached to Task Force Alpha in 1968 wore a wide variety of both tiger stripe and ERDL poplin uniforms cut along the lines of Vietnamese garments, with two upper breast pockets and an additional pen pocket on the left sleeve. The shirts were worn both to the end with the black open face buckle or worn un-tucked. A black on green "U.S. Marines" Corps strip is worn on the left side of the garment, and a green name strip with white lettering indicating the Headquarters Element would be worn on the right side in accordance with the Army uniform regulations. Worn on the left sleeve would be the early VNMC Brigade SSI, which was a black shield with the EGA in yellow and a red star. These came in a variety of either printed, handsewn, or silk woven. Pin-on rank would usually be worn on the collar points, although not always. Along with the tiger stripe is worn one of the Marine winter weight wool shirts, which featured a button closure and two additional breast pockets. These were worn for additional warmth during the cold monsoon seasons of Vietnam. Worn over his tiger stripe uniform is the M1955 Vest, Fragmentation, or "Flak" vest. This variation was produced in all nylon instead of earlier models over a cotton nylon blend and featured a double rope bridge, one on each of the shoulders, along with two lower cargo pockets made at the same time the vest was produced, instead of being added on, or retrofitted, later. These began appearing in late 1967. This creation also featured a strip of webbing with eyelets along the lower vest so that the M1910 type wire hangers could be attached. For example, here the Marine advisor has his black leather holster secured with his M1911A1 auto pistol.

Headgear (secured in his right hand) is the dark green Vietnamese beret, which would be worn French/Vietnamese style with

Close-up of Vietnamese Marine Corp beret with the metal Officer Cap Badge, which features a piece of red cloth behind the badge. (Mike Tucker Collection)

Additional close-up of two piece green beret with gold Officer Cap Badge. (Mike Tucker Collection)

U.S. Marine Corps advisor; Task Force Alpha, Saigon, TET offensive, 1968.

Full figure view shows early tiger stripe pattern with the large stripe pattern and flak vest. Note the dollar sign drawn above the small pocket on the left side.

Rear view of advisor gives a good look at the name printed on the back of the advisor's flak vest, "Dollar." Also note the early second model jungle boots and his black leather holster with his .45 pistol.

the badge over the right ear. The Vietnamese Marine Corps. beret device is based on the U.S.M.C. Eagle, Globe, and Anchor (EGA), but the country of Vietnam is substituted for the continents of North and South America. Also, a large red star is added to the background. The beret device is usually attached to a piece of red cloth, sometimes felt, or even red leather. This beret badge is fairly typical, being made out of tin, sometimes referred to as a "beer can" D.I. A gold device was for officers, and a silver device was intended for enlisted personnel.

Close-up of Marine Corp wool shirt.

Full figure view of advisor provides a good look at the black holster connected to his flak vest and early 1964 era jungle boots lacking the reinforcement strip on the side. Note the green Vietnamese Marine beret in his left hand.

Close-up showing the advisor's double rope bridge on his Marine Corp flak vest with the early style pockets, which features single exposed. Close-up shows the green the of the Marine Corps beret in advisor's left hand, along with the wool shirt underneath his large tigerstripe shirt. This also provides a good look at the black holster.

Left: *Vietnamese Marine wearing a uniform made from poplin ERDL material with the green Marine beret and the early Marine Brigade SSI on his left sleeve. Note also the U.S. black boots.*

USMC Advisor,
4th VNMC Battalion,
The Citadel Hue, TET
1968

"While Task Force Bravo and at least one other Vietnamese Marine Infantry Battalion attached to the Capital Military Command remained in Saigon, Task Force Alpha deployed to Hue and took part in the retaking of the Citadel in that city. From its initial commitment to II Corps, at the start of Tet, Task Force Alpha and its battalions had returned to Saigon to be in position to reinforce Task Force Bravo if needed. After losing operational control of two of its battalions, on 9 February the task force headquarters and the 1st Battalion departed Tan Son Nhut Airport by air for Phu Bai. By 14 February, the initial units were reinforced by two more battalions, the 4th and 5th. After some initial misunderstand-

Full figure view of Marine advisor to the 4th Vietnamese Marines, "Tet Offensive," Hue, 1968.

Full figure view of Marine advisor with the 4th South Vietnamese Marine Corps during the fighting in the Imperial city of Hue. Note the tigerstripe uniform, sometimes referred to as the "silver pattern," along with his M1955 flak vest and mismatched tiger stripe boonie hat.

ing, the commander of the 1st ARVN Division, General Ngo Quang Truong, assigned Task Force Alpha to clearing the western Citadel. Taking part in some of the heaviest fighting in the war, Task Force Alpha remained under the operational control of the first ARVN Division and in Hue until 27 March, when it relieved Task Force Bravo in Saigon. In the fighting for Hue, the Vietnamese Marine Task Force sustained casualties of nearly 90 dead and 350 wounded. All told, for the period 30 January through 27 March, which included the battles for both Saigon and Hue Vietnamese Marine reported killing over 1,300 of the Communists and capturing another 82 while detaining nearly 2,000 suspects. The entire Vietnamese Marine Corps suffered 128 killed, 588 wounded, and one missing in action (MIA)."[176]

This Marine Corps advisor to the South Vietnamese, 4th Marine Battalion is engaged in the heavy fighting to retake the Citadel in the Imperial City of Hue. Uniform is the silver pattern of tiger stripe, which features two upper chest pockets, each secured with two buttons. The matching trousers feature two rear pockets with a button fly, two large cargo pockets, and an additional cigarette pocket on the lower left leg, but lacked the front two slash pockets. Worn over his tiger stripes is the Marine Corps M1955, Armor Body Fragmentation Protective, which is an upgraded version of the M1951 Flak vest used during the Korean

Close-up of 2nd Battalion Vietnamese Marines; TRAU DIEN = Crazy Buffalo. Silk woven.

Full figure view of advisor wearing a boonie hat with his M1 helmet in his hand, along with his M-16 rifle. Note the 20 round magazine in the small pocket of his flak vest.

Full figure view provides a good look at the advisor's M-16 rifle. By this time selected Vietnamese units had access to these rifles. Hue, 1968.

War. Note the Marine advisor has placed an additional 20 round magazine for his M-16 rifle in the small pocket on the left side of the vest. This pocket was not intended to hold these magazines, but they fit and were often utilized this way. Underneath his flak vest he wears his M1961 load bearing equipment, which was unique to the Marine Corps. The M1961 equipment consisted of the pistol belt with male snaps, four M14 ammo pouches with female snaps that secured to the pistol belt, two aluminum canteens with covers, and an individual first aid kit known as the Jungle First Aid Kit.

Inside each of his M14 pouches the advisor has placed a single 20 round M-16 magazine. Although designed to carry a single M14 magazine, this advisor, although unable to procure the Army's universal pouches, has made due with his original Marine Corps gear. Additional 20 round magazines could be carried in either of his cargo pockets or in his rucksack. Headgear is a gold pattern tiger stripe boonie hat with an exceptionally wide brim. These

Right: *Vietnamese Marine in 1969 wearing a set of the Vietnamese Marine Corps "Sea Wave" tiger stripe pattern. Note he also wears the VNMC patch on the right pocket, along with a strip above his right pocket, believed to be a name strip. Note the VNMC beret badge being worn as a hat device on his matching boonie hat.*

Full figure view shows the advisor with his helmet off during a lull in the fighting around Hue. This also shows the Marines' early 1964 jungle boots with his socks pulled up and over the boots. Note, also, that he still uses the early metal canteens.

Rear view shows the advisor's metal canteens and WWII era jungle first aid kit.

could be purchased at any of the local tailor shops and garment shops in the area for almost nothing. In his left-hand he carries his M1 steel pot. Jungle boots are the second model, which lack the leather reinforcement along the top and additional ankle support on the side. Note how this advisor wears his socks pulled up and over the tops of his boots. Weapon is the M-16A1 rifle, which became common with the Vietnamese Marines just prior to the TET holiday.

USMC ADVISOR (NAG)
VIETNAMESE "JUNK FORCE,"
QUANG XUYEN DISTRICT TEAM
RUNG SAT SPECIAL ZONE (RSSZ)
1971

"By mid 1971, the main thrust of U.S. policy inside Vietnam was the advisory effort. Most Marines Advisors were concentrated within the framework of the Naval Advisory Group (NAG) under the commander, Naval Forces Vietnam, Rear Admiral Robert S. Salzer. This included the Marines who served their tours largely unheralded in the Mekong River Delta Region of MR4. The Rung Sat Special Zone (RSSZ) was divided into two districts, Can Gio on the east and Quang Xuyen on the west. Major James M. Tully

Close-up of gold officer's VNMC beret badge affixed to a green Vietnamese Marine Corps beret.

Full figure view of U.S. Marine Corps advisor to the South Vietnamese junk force along the Cambodian border, 1972.

Close-up showing the advisor in the ERDL ripstop shirt, black junk force beret, early style black leather shoulder holster for his .45 and M-79 grenade launcher.

was Commander Steward's assistant senior advisor and was specifically charged with tactical ground and air operations. Each district had a Marine Advisory team assigned to it. The Quang Xuyen district team was headed by Captain Ronald S. Neubauer. Captain Neubauer, a lean, red-headed Marine from Norwalk, Connecticut, was far removed from the pageantry of his previous duty station. While assigned to Marine barracks, Washington, DC, he had served as parade adjutant and social aid at the White House."[177] "While Captain David W. Blizzard headed the team in the Can Gio District. Daily, Vietnamese and Americans shared food, hardships, and work as they patrolled the waterways of the Rung Sat. To cover their districts, each Captain was assisted by six enlisted advisors and one Navy medical corpsman. The enlisted Marines, on occasion, were called upon to advise Vietnamese officers, while the Navy Corpsmen were often required to perform functions to or to give treatment that normally would be expected of a physician. The enlisted Marines were often communications, intelligence, and engineer specialists. They assisted in all forms of military and civil operations, but counterinsurgency effort was their forte. Each man on the team worked closely with

Members of the 6th South Vietnamese Marine Corps at the victory celebration at Quang Tri in 1972 following the defeat of the North Vietnamese Army during the Easter Offence, April 1972. The Vietnamese Marines wear both the Vietnamese Marine Corps "Sea Wave" pattern tiger stripe and the ARVN Ranger camouflage pattern uniforms. Note the "33" EXPORT beer.

his counterpart in an effort to develop a comprehensive program to pacify the entire Long Tau shipping channel."[178]

"Because of the importance of this network to the economies of both Vietnam and Cambodia, it was vital to keep these highways open. By October 1971, enemy activity had been brought under control. Combined U.S. Marines, U.S. Navy, and Army advisors, under the supervision of the senior advisor, RSSZ, U.S. Navy Commander Douglas A. Stewart, worked with local forces in operations designed to keep the enemy off-balance. At the same time, they assisted the civilians in reconstruction, public health, education, and other aspects of nation building. With some sense of security, the fishermen, the wood cutters, and farmers once again were following the roles of there forefathers as an ominous peace settled over the Rung Sat, the Forest of Death."[179]

"In such an undertaking, each meter of river was as vital as the next. Friendly Forces had to be as mobile and as flexible as possible in order to counter the Viet Cong threat. Operations ranging from multi-company helicopter and water board assaults, down to squad size interdiction missions were carried out repeatedly and with such effectiveness that any activity had been reduced to practically nothing. Advisors accompanied the Vietnamese on operations, assisting the commander on the ground in talking with the command and Communications helicopter, providing aerial observation, close air support, naval gunfire support, troop lifts, and medical evacuations."

The Mekong DELTA TAC ZONE:

"To the Northwest, along the Cambodian border, things had not been quiet. Late in 1970 the North Vietnamese and Viet Cong had made efforts to isolate Cambodia's Phnom Penh, the capital of the Khmer Republic, by closing all lines of communication, including the vital Mekong River between Tan Chau, Vietnam, and the capital. Early in 1971, the oil tanker Mekong was sunk while transporting petroleum upstream. After this attack the Cambodian government asked the United States and South Vietnam to provide protection for the convoys between Tan Chau and Phnom Penh. This was not to be an easy task, for convoy protection required close cooperation among nine military services of the three nations."[180]

Carried by this Marine Advisor to the "Junk Force" in 1971 is the M-79, affectionately called the "blooper" for the sound it made when the round left the chamber. Springfield Armory had begun development of this 40 x 46 mm SR weapon in the early 1950s. A basic goal of the weapon was the replacement of the standard rifle fired grenade, and 60 mm mortar at the squad and platoon level. The M79 could cover the area between the longest range of the hand grenade (30-40 meters) and the middle range of the 60 mm mortar (300-400 meters). In the hands of a skilled grenadier, the M79 was highly accurate to 200 meters, and worked well in the jungle and in building to building of urban warfare. The M79's basic combat round, the M-406, was a 170-grain high-explosive fragmentation grenade at a velocity of 76 m/s to a maximum range of 400 meters. Development of the grenade-type warhead for the weapon that would become the M79 came out of the research conducted at the U.S. Army Ballistic Research Laboratories (BRL) at Aberdeen Proving Ground.

It is a misconception that Vietnamese Marines and their USMC Advisors wore only the VNMC pattern tiger stripe uniform. Here we have an American Marine Advisor wearing a modified short sleeve ERDL shirt and matching rip-stop trousers. Worn above the left pocket is a U.S. MARINE tape. Worn slung across his chest is a U.S. made shoulder holster for his M1911A1 .45 auto pistol, worn for both a sign of authority and as a side arm. headgear is the black "Junk Force" beret with the twin sails gold officer device. Boots are the final model jungle boots with the Panama soles.

NAVAL ADVISORY GROUP; QUANG TRI PROVINCE; VNMC BRIGADE 147; MAI LOC COMBAT BASE NAVAL CORPSMAN. "EASTER OFFENCE" 1972

"Both the 147th and the 258th VNMC Brigades were under the command of the 3d ARVN Division. The 147 at this time was commanded by Major Jim R. Joy, who along with his Vietnamese Marine counterpart, Lieutenant Colonel Nguyen Nang Bao, were responsible for the western segment of the 3d ARVN Division's area of operation. In the VNMC Brigade 147th area were outposts at Nui Ba Ho and Sarge held by the 4th VNMC Battalions, along with two companies of the 8th VNMC Battalion operating in the vicinity of Fire Support Base Holcomb, forming the western flank of the defensive arc."

Full figure shot of U.S. Navy Corpsman attached to the MAU in 1971.

Full figure shot showing the front view of U.S. Navy Corpsman wearing the ARVN Ranger pattern in the lightweight poplin material, 1971.

Full figure shot of the rear of advisor provides a good close-up of the advisor's U.S. Army medic's bag and the USMC jungle first aid kit.

Many of the South Vietnamese Marine Officers had been trained in the U.S. and spoke English. They had been in combat for years and often did not feel the need for an American Advisor. Because of this, the American Marines often felt more like a fire support coordinator than an advisor. Because of this the senior USMC advisor had plans to pull the advisors from the battalion level and consolidate them at each brigade, thus forming a "liaison team" that could be responsive as needs arose.

This Naval Corpsman assigned to the USMC Advisor shows another example of Marine Advisors not wearing the distinct Vietnamese Marine Corps. tigerstripe pattern. The pattern worn is known as the ARVN Ranger pattern, and is a direct copy of the U.S. ERDL. The camouflage pattern is printed on lightweight fabric, and has the same bright green tones of the "Lowland" ERDL pattern. It was not uncommon to see VNMC members or even their advisors wearing the ARVN Ranger pattern this late in the war. Unlike the "typical" VNMC pattern tigerstripe shirt, this one lacks the shoulder reinforcement, shoulder loops, and the pen pocket. But it does feature the typical flush breast pocket, secured with two buttons. The matching ARVN Ranger pattern pants are printed on the same lightweight material, and feature two rear pockets, two front "slash" pockets, but lack the two cargo pockets, and feature a button fly. The black web trouser belt is worn

Close-up showing the Navy Corpsman's 1st model jungle boots with the leather reinforcement along the top of the boots.

Close-up showing the top of the 1st model boots. This is the true 1st model production jungle boots!

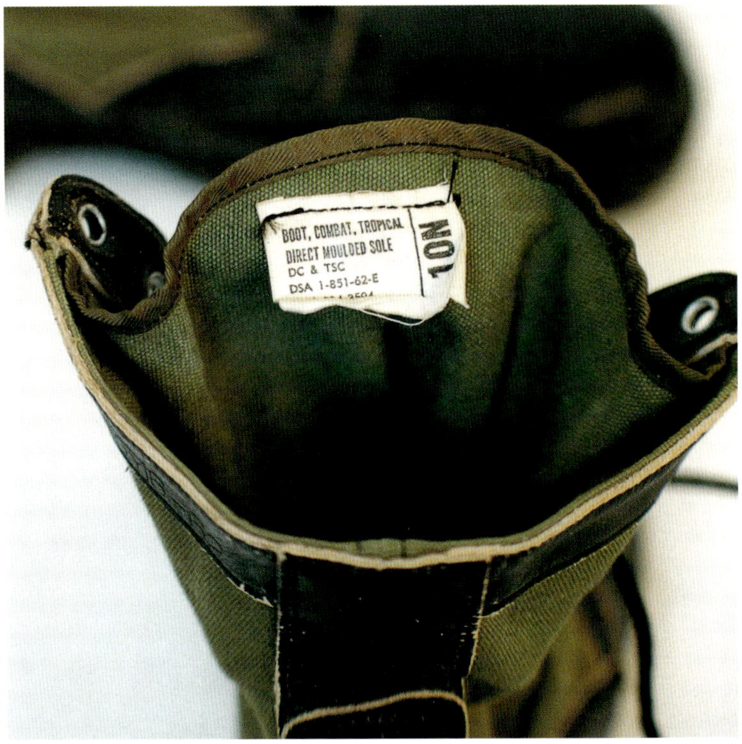

Close-up showing the white cotton tag in the 1st model jungle boots. Note the date in the code—DSA 1-851-62-E—the date is 1962. Note the leather strip up the rear of the boot.

with a flat black "open face" belt buckle. Around his neck he wears the ever popular green "drive on" rag. On his wrist he wears a Timex® wristwatch, probably purchased at the Army PX or Navy BX.

LBE consists of basic U.S. Army web gear. A total of four Universal pouches are worn, with two Marine M1943 canteens and covers. Worn on the rear of the pistol belt is the Marine Corps. "jungle" first aid kit, a standard Marine item, especially when working with the Vietnamese. Rucksack is the U.S. Army Medical Backpack, which were being issued to the Naval Corpsman by this phase of the war. The medical field pack is worn correctly as a backpack, and was capable of carrying enough bandages for an Infantry platoon. Most Corpsman carried triangular bandages, Vaseline gauze, numerous dressings, field medical cards, anti-malaria tablets, water purification tablets, and bacitracin and tetraccin ointments, along with numerous cough longezes, numerous large pressure dressings, bandages, saline solutions, and aspirin. There was no SOP for packing the medical kit; the Corpsman filled the pack with what he deemed necessary for the allotted time they would be out. Jungle boots, worn here, are the true 1st model DMS jungle boots, which feature the leather strip along the top, which has since been replaced by nylon, and a section of leather running up vertically along the boot's rear. (See Close up) The advisor/Corpsman wears a boonie hat, which is in the true VNMC tigerstripe pattern, which features a short brim and two eyelet/ventilated screens to aid in ventilation. Attached to the front of the boonie hat are two blossoms showing his rank as Dai Uy, Lieutenant, JG. Weapon is an M-16A1 with a 20 round magazine.

Close-up of VNMC patch.

U.S.M.C. Advisor,
6th Vietnamese Marine Corps. (VNMC)
Quang Tri Province.
1972 "Easter Offence"

On 5 April, General Giai had alerted the 6th VNMC at Ai Tu to prepare to move to the vicinity of Dong Ha at first light the next day. Major Do Huu Tung, the Battalion Commander, after a reconnaissance to Dong Ha, began briefing his unit on the move north. Assisting in this were his advisors, Major William R. Warren and Captain William D. Wischmeyer. By 1300, however, the plans had been changed; the 6th VNMC was going to Fire Sup-

Full figure shot of USMC advisor to the SVN Marine Corps in 1972 during the Easter Offence, 1972.

Full rear shot of Marine advisor wearing the "Sea Wave" pattern. Note the ARVN rucksack, along with the M17 protective mask.

Full figure front view of the Marine advisor provides a good look at the advisor wearing the U.S. Army's M69 flak vest, 1972.

port Base "Pedro" instead, and the 1st VNMC was moved back to Ai Tu to assume a portion of the perimeter defense.

This Marine Corps advisor wears a set of the Vietnamese Marine Corps tiger stripe known as the "Sea Wave" pattern. This pattern was developed and issued to the Vietnamese Marines sometime in the mid-'60s after switching from the heavier material and larger stripe pattern. The shirt features two breast pockets, reinforced shoulders, and epaulets. On the left sleeve just below his shoulder is the Marine Corps patch, and a rather large pen pocket; additional pen pockets have also been added on each side underneath the armpit. This unique tiger stripe pattern was always printed on lightweight material. Although obstructed by his flak vest, the U.S. advisor wears a subdued U.S. Marines strip and a white on green nametape, indicating he is a member of the headquarters element. His equivalent Vietnamese rank would usually be worn on his right pocket flap as evident by many photos. Worn on the right pocket would be the circular, silk woven Vietnamese Marine Corps patch which read "Danh Du To Quoc" (Honor and Country); this was the same as the USMC stencil worn by U.S. Marines on their left pocket. Additional battalion patches would typically be worn on the right sleeve.

Body armor worn here is the M69 Fragmentation Protective Vest with three quarter collar, which was an upgraded version of the M1952 Army vest. It was not uncommon to see Marine Corps

Close-up VNM SOI BIEN, or "Sea Wolves."

Close-up showing the Marine advisor's locally made Marine Corps cover in the traditional "Sea Wave" pattern. This also provides a good look at the advisor's U.S. Army flak vest and his "deuce gear" worn over the flak vest.

Vietnamese Marine Officer wearing a set of the ARVN Ranger pattern with a matching ARVN Ranger pattern boonie hat, along with a SVN Marine Corp beret device worn as a cap badge. Circa 1971-72.

Vietnamese Marine Corps Officer wearing the "Sea Wave" pattern with two white strips and the VNMC Corp patch on his right pocket. Note the VNMC Eagle Globe and Anchor worn on his tigerstripe fatigue cap.

advisors wearing Army flak vests and wearing the army web gear, also. Web gear consists of a M1961 Marine Corps. pistol belt with two universal pouches, a single M26 hand grenade, two M1943 canteen covers, and a WWII russet brown holster for his M1911 A1 .45 pistol. On his back he carries his personal items secured in an ARVN rucksack with an M17 protective mask secured to the outside. Weapon is the M-16 A1 rifle. Headgear is the M1 steel pot with the Mitchell camouflage cover.

Close-up of Vietnamese made tiger stripe short brim boonie hat in the "Sea Wave" pattern. Note at the top you can see the owners name, Holly.

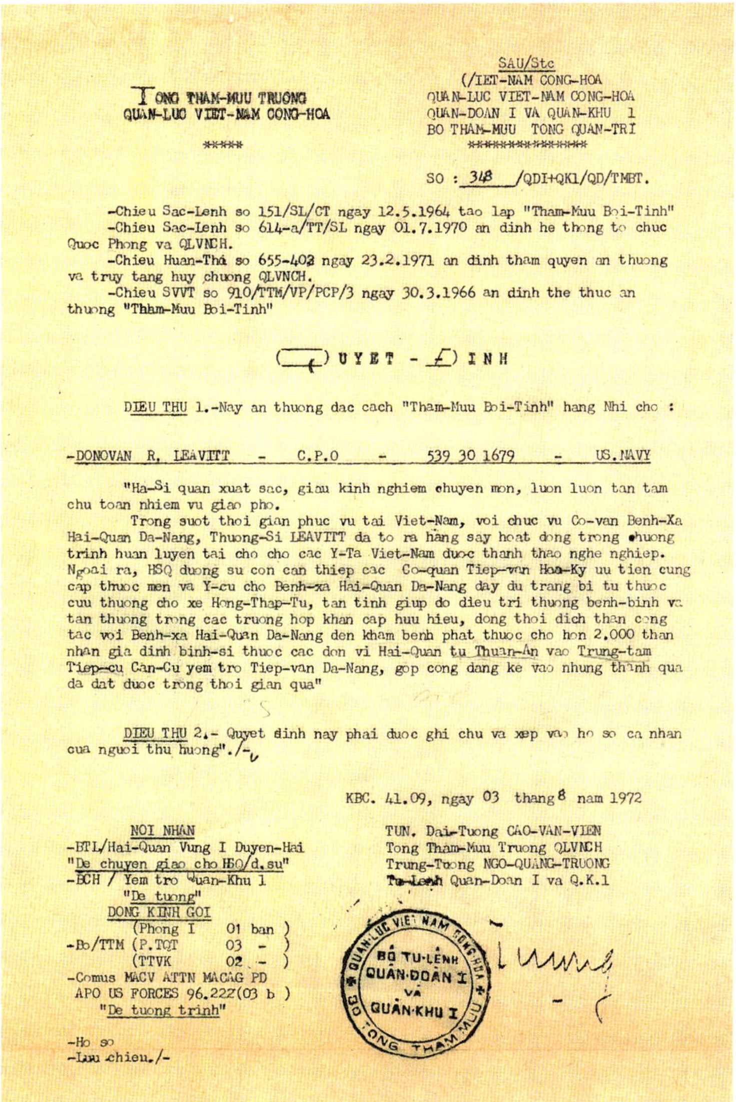

SAU/Stc

TONG THAM-MUU TRUONG
QUAN-LUC VIET-NAM CONG-HOA
*****

(/IET-NAM CONG-HOA
QUAN-LUC VIET-NAM CONG-HOA
QUAN-DOAN I VA QUAN-KHU 1
BO THAM-MUU TONG QUAN-TRI
****

SO : 348 /QDI+QK1/QD/TMBT.

-Chieu Sac-Lenh so 151/SL/CT ngay 12.5.1964 tao lap "Tham-Muu Boi-Tinh"
-Chieu Sac-Lenh so 614-a/TT/SL ngay 01.7.1970 an dinh he thong to chuc Quoc Phong va QLVNCH.
-Chieu Huan-Thi so 655-402 ngay 23.2.1971 an dinh tham quyen an thuong va truy tang huy chuong QLVNCH.
-Chieu SVVT so 910/TTM/VP/PCP/3 ngay 30.3.1966 an dinh the thuc an thuong "Tham-Muu Boi-Tinh"

QUYET - DINH

DIEU THU 1.-Nay an thuong dac cach "Tham-Muu Boi-Tinh" hang Nhi cho :

-DONOVAN R. LEAVITT - C.P.O - 539 30 1679 - US.NAVY

"Ha-Si quan xuat sac, giau kinh nghiem chuyen mon, luon luon tan tam chu toan nhiem vu giao pho.

Trong suot thoi gian phuc vu tai Viet-Nam, voi chuc vu Co-van Benh-Xa Hai-Quan Da-Nang, Thuong-Si LEAVITT da to ra hang say hoat dong trong chuong trinh huan luyen tai cho cho cac Y-Ta Viet-Nam duoc thanh thao nghe nghiep. Ngoai ra, HSQ duong su con can thiep cac Co-quan Tiep-van Hoa-Ky uu tien cung cap thuoc men va Y-cu cho Benh-xa Hai-Quan Da-Nang day du trang bi tu thuoc cuu thuong cho xe Hong-Thap-Tu, tan tinh giup do dieu tri thuong benh-binh va tan thuong trong cac truong hop khan cap huu hieu, dong thoi dich than cong tac voi Benh-xa Hai-Quan Da-Nang den kham benh phat thuoc cho hon 2.000 than nhan gia dinh binh-si thuoc cac don vi Hai-Quan tu Thuan-An vao Trung-tam Tiep-cu Can-Cu yem tro Tiep-van Da-Nang, gop cong dang ke vao nhung thanh qua da dat duoc trong thoi gian qua"

DIEU THU 2.- Quyet dinh nay phai duoc ghi chu va xep vao ho so ca nhan cua nguoi thu huong"./-

KBC. 41.09, ngay 03 thang 8 nam 1972

NOI NHAN
-BTL/Hai-Quan Vung I Duyen-Hai
"De chuyen giao cho HSQ/d.su"
-BCH / Yem tro Quan-Khu 1
"De tuong"
DONG KINH GOI
(Phong I 01 ban)
-Bo/TTM (P.TQT 03 -)
(TTVK 02 -)
-Comus MACV ATTN MACAG PD
APO US FORCES 96.222(03 b)
"De tuong trinh"

-Ho so
-Luu chieu./-

TUN. Dai-Tuong CAO-VAN-VIEN
Tong Tham-Muu Truong QLVNCH
Trung-Tuong NGO-QUANG-TRUONG
Tu-Lenh Quan-Doan I va Q.K.1

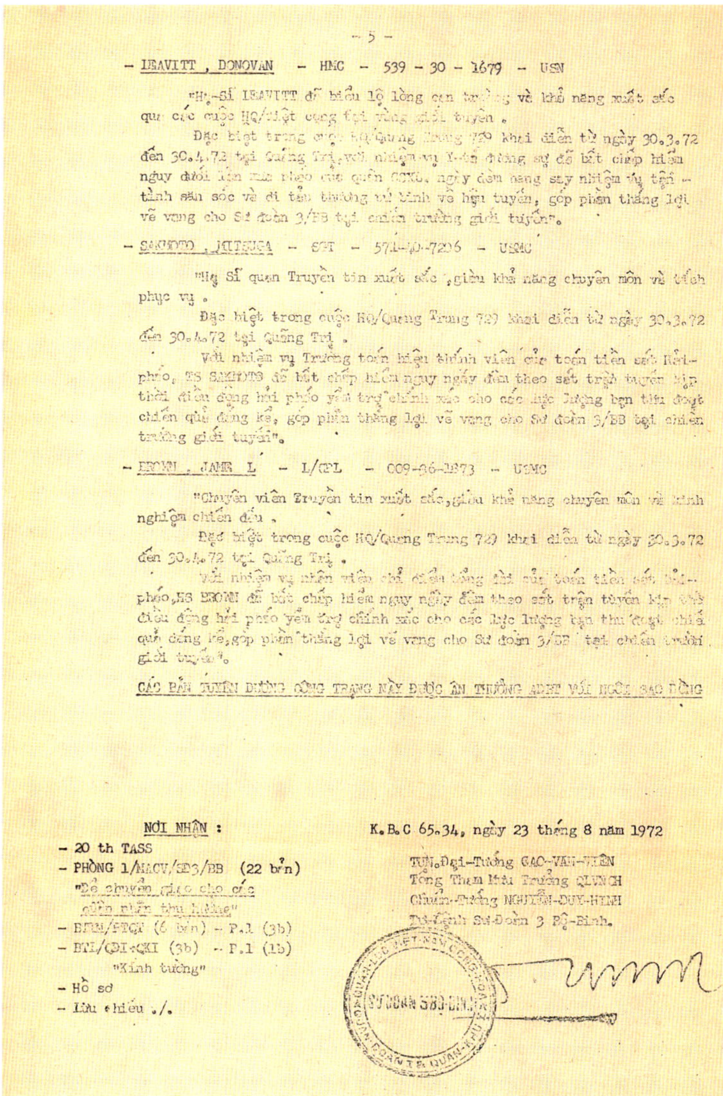

- 5 -

- LEAVITT , DONOVAN - HMC - 539 - 30 - 1679 - USN

"Hạ-Sĩ LEAVITT đã biểu lộ lòng can trường và khả năng xuất sắc qua các cuộc HQ/tiệt cọng tại vùng giới tuyến .

Đặc biệt trong cuộc HQ/Quang Trung 729 khai diễn từ ngày 30.3.72 đến 30.4.72 tại Quảng Trị, với nhiệm vụ [illegible] đã bất chấp hiểm nguy dưới [illegible] của quân CSBV, ngày đêm hăng say nhiệm vụ tận tình săn sóc và di tản thương bệnh binh về hậu tuyến, góp phần thắng lợi vẻ vang cho Sư đoàn 3/BB tại chiến trường giới tuyến".

- SAKIMOTO , MITSUGA - SGT - 571-40-7206 - USMC

"Hạ Sĩ quan Truyền tin xuất sắc, giàu khả năng chuyên môn và tích phục vụ .

Đặc biệt trong cuộc HQ/Quang Trung 729 khai diễn từ ngày 30.3.72 đến 30.4.72 tại Quảng Trị .

Với nhiệm vụ Trưởng toán hiệu thính viên của toán tiền sát Hải-pháo, TS SAKIMOTO đã bất chấp hiểm nguy ngày đêm theo sát trận tuyến kịp thời điều động hải pháo yểm trợ chính xác cho các lực lượng bạn thu đoạt chiến quả đáng kể, góp phần thắng lợi vẻ vang cho Sư đoàn 3/BB tại chiến trường giới tuyến".

- BROWN , JAME L - L/CPL - 009-36-1873 - USMC

"Chuyên viên Truyền tin xuất sắc, giàu khả năng chuyên môn và kinh nghiệm chiến đấu .

Đặc biệt trong cuộc HQ/Quang Trung 729 khai diễn từ ngày 30.3.72 đến 30.4.72 tại Quảng Trị .

Với nhiệm vụ nhân viên chỉ điểm [illegible] của toán tiền sát hải-pháo, HS BROWN đã bất chấp hiểm nguy ngày đêm theo sát trận tuyến kịp thời điều động hải pháo yểm trợ chính xác cho các lực lượng bạn thu đoạt chiến quả đáng kể, góp phần thắng lợi vẻ vang cho Sư đoàn 3/BB tại chiến trường giới tuyến".

CÁC BẢN TUYÊN DƯƠNG CÔNG TRẠNG NÀY ĐƯỢC ÂN THƯỞNG ADHT VỚI NGÔI SAO ĐỒNG

NƠI NHẬN :
- 20 th TASS
- PHÒNG 1/MACV/SD3/BB (22 bản)
"Để chuyển giao cho các quân nhân thụ hưởng"
- BTTM/PTQT (6 bản) - P.1 (3b)
- BTL/QĐI-QKI (3b) - P.1 (1b)
"Kính tường"
- Hồ sơ
- Lưu chiếu ./.

K.B.C 65.34, ngày 23 tháng 8 năm 1972

TUN.Đại-Tướng CAO-VĂN-VIÊN
Tổng Tham Mưu Trưởng QLVNCH
Chuẩn-Tướng NGUYỄN-DUY-HINH
Tư-Lệnh Sư-Đoàn 3 Bộ-Binh.

U.S. Naval advisor's Vietnamese orders for the Cross of Gallantry.

Typical U.S. Marine Corps advisor to the Vietnamese Marine Corps tiger stripe uniform and matching tiger stripe boonie hat in the same "Sea Wave" pattern. (Mike Tucker collection)

Additional photo of Capt. Hernandez's Vietnamese tiger stripe uniform, which provides a good look at the direct embroidered U.S. Captain bars on the collars, along with a direct embroidered Vietnamese Captain rank on the pocket flap. (Mike Tucker collection)

Close-up of matching tiger stripe boonie hat that once belonged to Captain Hernandez. Note the pin-on Captain bars that are Vietnamese made. (Mike Tucker collection)

Additional close-up provides a good look at the boonie hat's construction and the pin-on 0-3 rank. This pattern of tiger stripe is one of the rarest and most sought-after. (Mike Tucker collection)

Photo of the legendary Col. Dorsey, Senior Marine advisor's uniform. (Mike Tucker collection)

Additional photo showing Col. Dorsey's Vietnamese Marine Corps tiger stripe uniform. Note the final model VNMC shoulder sleeve insignia with the typical pen pocket, reinforced shoulders, and shoulder loops. Also note the U.S. Marines strip and basic parachute wings. (Mike Tucker collection)

Close-up showing the U.S. Marines strip and hand sewn U.S. airborne wings of Col. Dorsey's uniform. Note the Marine Corps strip is machine made. (Mike Tucker collection)

Close-up showing the white on green name tape of Col. Dorsey. (Mike Tucker collection)

Additional close-up provides a good look at his direct embroidered Colonel rank on the collars, along with his Vietnamese Marine Corps pocket patch, which is the equivalent to the U.S. Marine course stencil. (Mike Tucker collection)

Photo provides a good look at the advisor's right side. Note the Vietnamese made name tape that was once a forest green with white lettering, indicating he is a member of the Headquarters Company. Most USMC advisors wore this colored name tape. Just below that on the pocket flap is a hand sewn Vietnamese equivalent (05) Colonel rank, and the Vietnamese Marine Corps pocket patch. (Mike Tucker collection)

Close-up showing the direct embroidered Col. rank of Col. Dorsey's uniform. This provides a good look at the quality and craftsmanship that went into the uniforms. (Mike Tucker collection)

Close-up of the direct embroidered Colonel rank on the collar of Col. Dorsey's tigerstripe uniform. This also provides a good look at the shoulder loops and reinforcements of the uniform. (Mike Tucker collection)

7

Personal Items

WWII era leather Jungle boots, featuring the double buckle and canvas uppers. These were issued to advisors all the way up to 1965.

Additional photo of WWII "double buckle" jungle boots, which were produced too late to see combat in the Pacific. Some soldiers were issued these just in time for Korea. These were known as the "Okinawa Boots" by Special Forces advisors assigned to the island.

Close-up showing the russet brown leather lowers of the WWII jungle boot, the two eye holes for ventilation, and the canvas uppers.

Close-up showing the date and size on the leather buckle. Although a little hard to read, the boots were produced in 1951.

Close-up of the cleated sole, which gave satisfactory traction, but had a problem with bonding failure, and would separate from the boot in hot, wet environments.

DMS tropical combat boot with Vibram soles. This is the true first model DMS jungle boot first produced in large quantities in 1962. They feature the leather strip across the top of a nylon duck upper, and additional leather band up the Achilles tendon.

The second model DMS jungle boots without the leather strip across the top or the back. They feature the Vibram sole and lacked the ankle reinforcement. These were first produced in great quantities in 1964.

Close-up showing date on second model jungle boots—10 R 665—which shows these boots are 10 regular and were produced in June 1965.

Additional close-up of second model boots.

Close-up of white cotton tag in second model boots. Later models would do away with this tag.

Additional close-up of white cotton tag in the Boot, Combat, Tropical, Direct Molded Sole. If you look close, you can see the date is 1964.

Close-up of vertical stitch which is found on every pair of jungle boots prior to 1975. After the war the vertical stitch would be done away with. Other than this stitch, postwar jungle boots are virtually impossible to recognize from wartime production.

4th and final model DMS jungle boots. They feature ankle reinforcements and the Panama sole. These boots would remain unchanged for the next 20 years.

Close-up of printed label of the front of the jungle boot. It reads 11N, showing its 11 narrow, and 6 67, showing these were produced in June 1967. Note also the spike protective.

Close-up showing the all leather McNamara boot, named after Secretary of Defense Robert McNamara.

Close-up showing the DMS all leather black combat boots. These came into service around January 1967. These were worn by aviators and in other fields where the nylon jungle boots were not worn because of the risk of fire hazard.

Close-up of typical Vietnamese made cowboy hat. The tailor used material from a U.S. parachute for the hat.

Close-up of another style of the Vietnamese made cowboy hat. This one, too, is in the parachute material camouflage, but is styled in the French design.

Typical Vietnamese made tan cowboy hat with the addition of a blue and white silk Vietnam arc. These were bought at the local tailor shops anywhere in Vietnam.

Vietnamese made tiger stripe cowboy hat, which features two snaps, one on each side, and a blue and white chin strap, similar to ones on kid's cowboy hats of the 1950s and 60s.

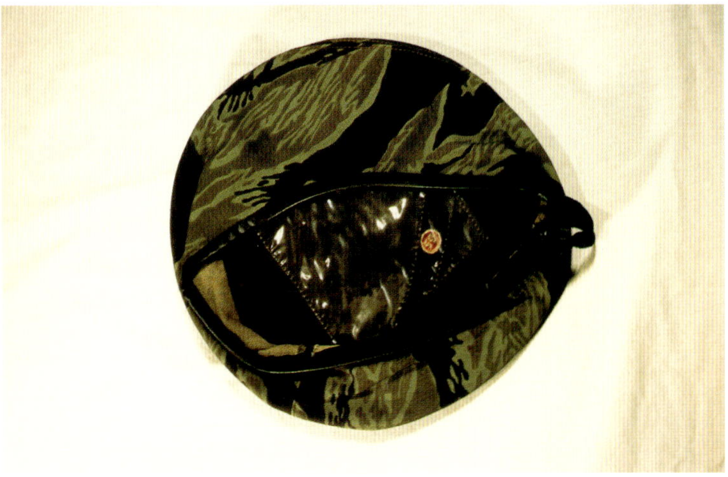

Vietnamese made tiger stripe beret in the traditional "Sea Wave" Vietnamese Marine Corps pattern.

Locally made sewing kit.

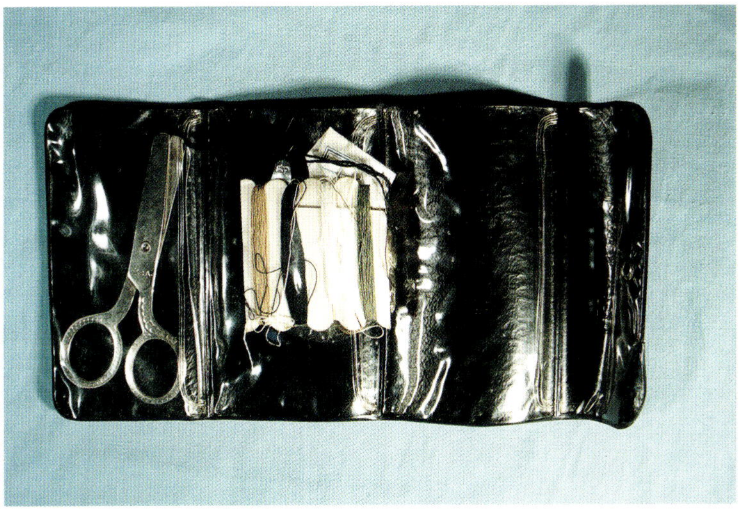

Contents of the locally produced sewing kit. Inside would be the usual needles, thread, and scissors.

Close-up of U.S. made sewing kit.

Contents of U.S. made sewing kit minus the needles. If you look closely, you can see the date, 1962.

Locally made Coconut rice bowl, chop sticks, and rice spoon. The spoon is made from aircraft aluminum.

Two containers of sunburn preventive preparation, i.e. sunscreen. Both are dated 1968, but were produced in December 1967.

Close-up of 1966 dated sunburn preventive preparation cream, or sunscreen.

A small container of anti-fungal foot powder.

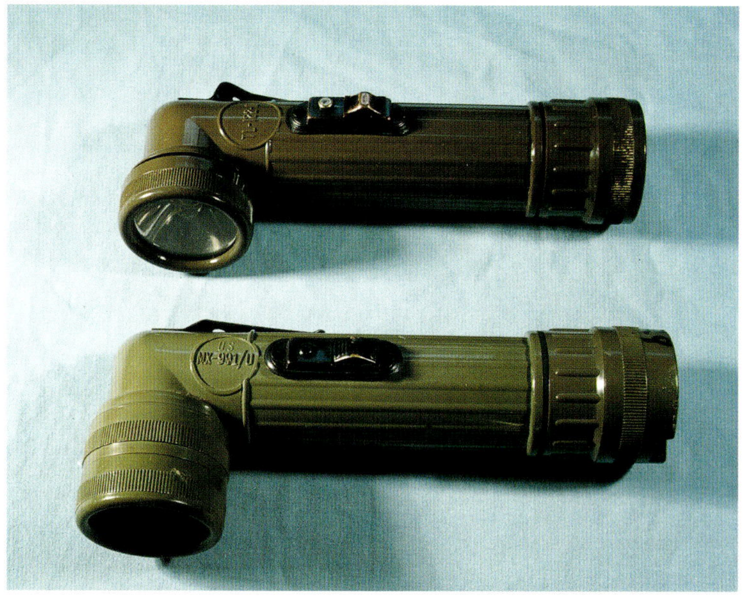

Close-up showing the differences of the WWII era angle head flashlight (top) and the Vietnam War era angle head flashlight (bottom). Note the differences are very slight.

Close-up showing the pink safety flashlight.

Kodak Instamatic camera, circa 1964-65.

Close-up showing green issue notebook, Memoranda.

Close-up showing green issue Army notebook on the left and the soldier's notebook, which could be purchased in the post exchange on the right.

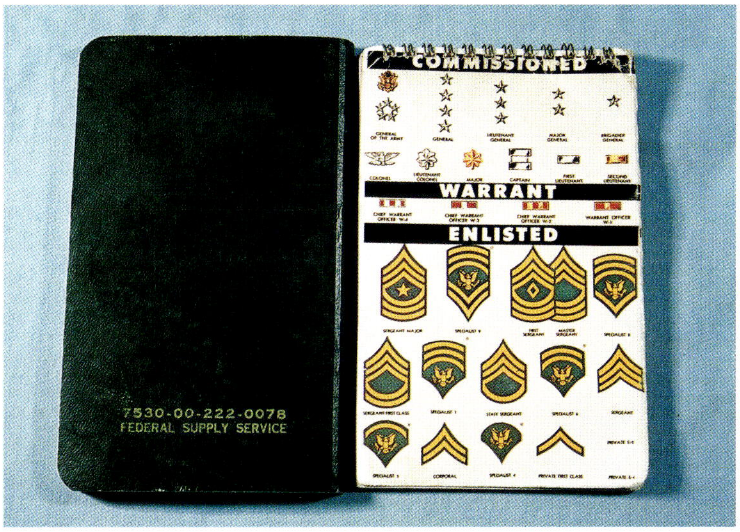

Close-up showing the back of both notebooks. The private purchase one on the right was full of helpful hints and useful information.

Close-up showing the Country Study and Station Report, Republic of Vietnam. The military Assistance Institute, Department of Defense, put out this manual. The book is filled with useful information. (see chapter 1, MAAG)

Close-up of small pamphlet put out by Department of State entitled Fact Sheet. It talks mainly about the current political and economic conditions of the Republic of Vietnam in 1963. These were given to most advisors prior to leaving for Vietnam.

Close-up of issue New Testament Bible.

Close-up of 1965 dated technical manual; TM 11-5820-398-12, for the radio set A. N./PRC 25. (Courtesy A. J. Allen)

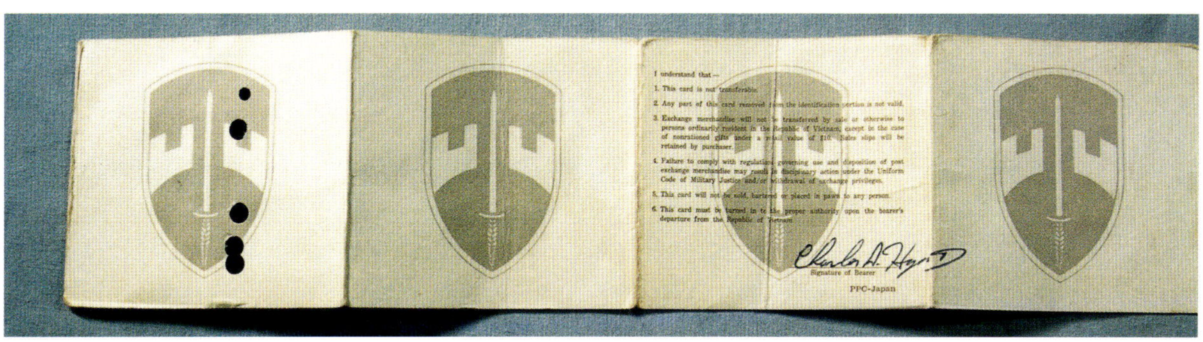

Close-up of MACV ration card dated 1972.

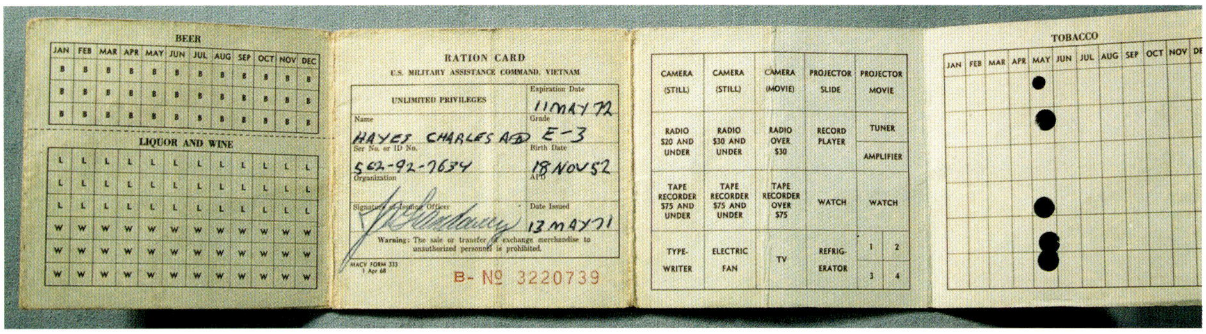

Inside of MACV ration card shows what the bearer of the card was entitled to each month.

Close-up of a pocket guide to Vietnam. These told of the culture of the people of Republic of Vietnam, along with the rank structure of the armed forces of Vietnam and useful phrases for getting by in the cities.

Close-up of South Vietnamese propaganda leaflets dropped into enemy held territory to try to convince the VC to come over to the "winning" side.

Much sought-after 10 dollar MPC, which featured a Special Forces soldier on one side and a U.S. M48 tank on the reverse.

Close-up of Vietnamese letter with two Republic of Vietnam stamps.

Close-up of military payment certificate (MPC) series 661. Military personnel would exchange U.S. currency, i.e. green backs, for MPC, which helped keep American dollars off the black market.

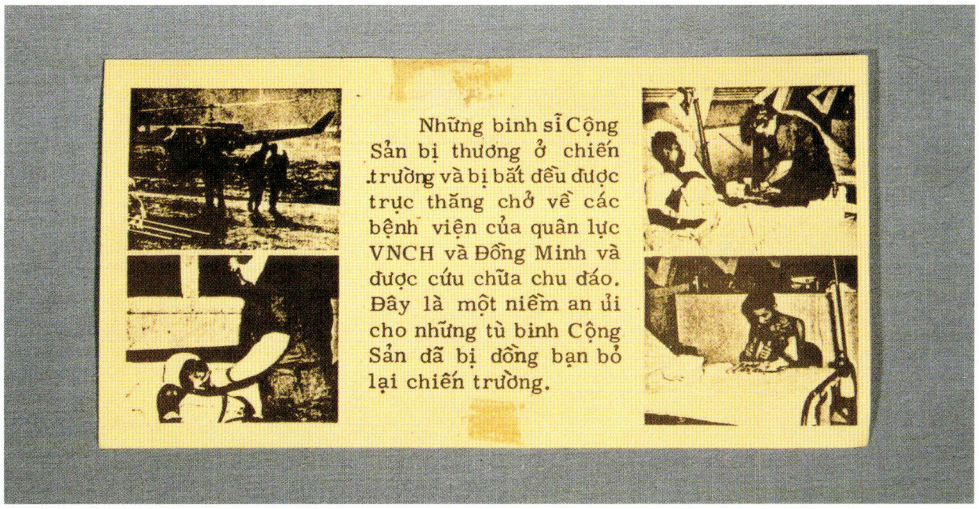

Back side of South Vietnamese propaganda leaflet.

Laotian or Cambodian made black teak wood decorative plate shows an elephant pushing a log. With bamboo. Done in 14K gold.

Front side of Vietnamese service medal.

Back side showing the medal is actually Vietnamese made; it reads CHIEN-DICH, BOI-TINH.

Several Vietnamese made medals.

Vietnamese made plastic political button with a picture of president Diem on it. Circa 1961.

Numerous personal effects, which would be found at the advisors' compound.

Bibliography

Cutler, Thomas J. Brown Water, Black Beret. Annapolis, Maryland: Naval Institute Press, 1988.

Lyles, Kevin.Vietnam: U.S. Uniforms in Colour Photographs. London, England: Windrow & Greene., 1992.

Lyles, Kevin. U.S. Infantry Vietnam. Hong Kong: Concord Ltd., 1996

Lyles, Kevin. U.S. Infantry Vietnam. Hong Kong: Concord Ltd., 1994

Martin, CSM Mike, LTC Valentine, M. The Black Tigers. Indiana: Walsworth Publishing Company., 1998

Rottman, Gordon L. U.S. Army Rangers and LRRP Units, 1942-87. London, England: Osprey Publishing Ltd., 1987.

Rottman, Gordon Panama, 1989-90. London, England: Osprey Publishing Ltd., 1991.

Stanton, Shelby. U.S. Uniforms of the Vietnam War. Harrisburg, Pa: Stackpole Books., 1989.

Stanton, Shelby. U.S. Uniforms of the Cold War, 1948-1973. Mechanicsburg, Pa: Stackpole Books., 1994

Stanton, Shelby.The Rise and Fall of an American Army. U.S. Ground Forces in Vietnam, 1965-1973.Novato, Ca: Presidio Press.,1985.

Stanton, Shelby. Vietnam Order of Battle. New York: Galahand Books.,1986.

Stanton, Shelby. Rangers at War. New York: Ivy Books., 1992.

Sutherland, Ian D. 1952/1982 Special Forces of the United States Army. San Jose: Bender Publishing., 1982

Notes

[1] Lyles, Kevin.*Vietnam: U.S. Uniforms in Colour Photographs.* London, England: Windrow & Greene., 1992. pp3
[2] Cutler, Thomas J. *Brown Water, Black Beret.* Annapolis, Maryland: Naval Institute Press, 1988.pp14
[3] Cutler, Thomas J. *Brown Water, Black Beret.* Annapolis, Maryland: Naval Institute Press, 1988. pp14
[4] Cutler, Thomas J. *Brown Water, Black Beret.* Annapolis, Maryland: Naval Institute Press, 1988. pp16
[5] Cutler, Thomas J. *Brown Water, Black Beret.* Annapolis, Maryland: Naval Institute Press, 1988.pp17
[6] CUTLER, PP19 Cutler, Thomas J. *Brown Water, Black Beret.* Annapolis, Maryland: Naval Institute Press, 1988.pp19
Specter, Ronald H. *Advice and Support. The Early Years; The US Army In Vietnam.* Washington, D.C., Center of Military History United States Army, 1983. Pp291
[7] Specter, Ronald H. *Advice and Support. The Early Years; The US Army In Vietnam.* Washington, D.C., Center of Military History United States Army, 1983. Pp291
[8] Specter, Ronald H. *Advice and Support. The Early Years; The US Army In Vietnam.* Washington, D.C., Center of Military History United States Army, 1983. Pp292
[9] Specter, Ronald H. *Advice and Support. The Early Years; The US Army In Vietnam.* Washington, D.C., Center of Military History United States Army, 1983. Pp293
[10] Specter, Ronald H. *Advice and Support. The Early Years; The US Army In Vietnam.* Washington, D.C., Center of Military History United States Army, 1983. Pp293
[11] Specter, Ronald H. *Advice and Support. The Early Years; The US Army In Vietnam.* Washington, D.C., Center of Military History United States Army, 1983. pp294
[12] Stanton, Shelby, *US Army Uniforms of the Cold War, 1948-1973*, (Mechanicsburg, PA, Stackpole Books, 1994), pp 113
[13] Stanton, Shelby, *US Army Uniforms of the Cold War1948-1973*, (Mechanicsburg, PA, Stackpole Books, 1994), pp 113-114
[14] Stanton, Shelby, *US Army Uniforms of the Cold War1948-1973*, (Mechanicsburg, PA, Stackpole Books, 1994), pp 116
[15] Stanton, Shelby, *US Army Uniforms of the Cold War1948-1973*, (Mechanicsburg, PA, Stackpole Books, 1994), pp 116-117
[16] Stanton, Shelby, *US Army Uniforms of the Cold War1948-1973*, (Mechanicsburg, PA, Stackpole Books, 1994), pp 117
[17] Stanton, Shelby, *US Army Uniforms of the Cold War1948-1973*, (Mechanicsburg, PA, Stackpole Books, 1994), pp 130
[18] Stanton, Shelby, *US Army Uniforms of the Cold War1948-1973*, (Mechanicsburg, PA, Stackpole Books, 1994), pp 80
[19] Stanton, Shelby, *US Army Uniforms of the Cold War1948-1973*, (Mechanicsburg, PA, Stackpole Books, 1994), pp 81
[20] Stanton, Shelby, *US Army Uniforms of the Cold War1948-1973*, (Mechanicsburg, PA, Stackpole Books, 1994), pp 85
[21] Stanton, Shelby, *US Army Uniforms of the Cold War1948-1973*, (Mechanicsburg, PA, Stackpole Books, 1994), pp 85
[22] Stanton, Shelby, *US Army Uniforms of the Cold War1948-1973*, (Mechanicsburg, PA, Stackpole Books, 1994), pp 85-86
[23] Stanton, Shelby, *US Army Uniforms of the Cold War1948-1973*, (Mechanicsburg, PA, Stackpole Books, 1994), pp 87
[24] Stanton, Shelby, *US Army Uniforms of the Cold War1948-1973*, (Mechanicsburg, PA, Stackpole Books, 1994), pp 87
[25] Stanton, Shelby, *US Army Uniforms of the Cold War1948-1973*, (Mechanicsburg, PA, Stackpole Books, 1994), pp108
[26] Stanton, Shelby, *US Army Uniforms of the Cold War1948-1973*, (Mechanicsburg, PA, Stackpole Books, 1994), pp108
[27] Stanton, Shelby, *US Army Uniforms of the Cold War1948-1973*, (Mechanicsburg, PA, Stackpole Books, 1994), pp108
[28] Stanton, Shelby, *US Army Uniforms of the Cold War1948-1973*, (Mechanicsburg, PA, Stackpole Books, 1994), pp 125
[29] Stanton, Shelby, *US Army Uniforms of the Cold War1948-1973*, (Mechanicsburg, PA, Stackpole Books, 1994), pp 125
[30] Stanton, Shelby, *US Army Uniforms of the Cold War1948-1973*, (Mechanicsburg, PA, Stackpole Books, 1994), pp 225
[31] Stanton, Shelby, *US Army Uniforms of the Cold War1948-1973*, (Mechanicsburg, PA, Stackpole Books, 1994), pp 126
[32] Stanton, Shelby, *US Army Uniforms of the Cold War1948-1973*, (Mechanicsburg, PA, Stackpole Books, 1994), pp 129-130
[33] Stanton, Shelby; *US Army Uniforms of the Vietnam War;* (Harrisburg, PA, Stackpole Books, 1988), pp 225
[34] Lyels, Kevin, Vietnam: *US Uniforms in Colour Photographs*, (London W1, Windrow & Greene, 1992), pp 7
[35] Lyels, Kevin, Vietnam: *US Uniforms in Colour Photographs*, (London W1, Windrow & Greene, 1992), pp 5
[36] Lyels, Kevin, Vietnam: *US Uniforms in Colour Photographs*, (London W1, Windrow & Greene, 1992), pp 5
[37] Stanton, Shelby, *US Army Uniforms of the Cold War1948-1973*, (Mechanicsburg, PA, Stackpole Books, 1994), pp 145-146
[38] Clarke, Jeffrey J., *Advice and support: The Final Years, The US Army In Vietnam,* (Washington, D.C. Center of Military History

United States Army, 1988), pp49

[39] Clarke, Jeffrey J., *Advice and support: The Final Years, The US Army In Vietnam,* (Washington, D.C. Center of Military History United States Army, 1988), pp49

[40] Clarke, Jeffrey J., *Advice and support: The Final Years, The US Army In Vietnam,* (Washington, D.C. Center of Military History United States Army, 1988), pp56-57

[41] Clarke, Jeffrey J., *Advice and support: The Final Years, The US Army In Vietnam,* (Washington, D.C. Center of Military History United States Army, 1988), pp57

[42] Clarke, Jeffrey J., *Advice and support: The Final Years, The US Army In Vietnam,* (Washington, D.C. Center of Military History United States Army, 1988), pp55

[43] Clarke, Jeffrey J., *Advice and support: The Final Years, The US Army In Vietnam,* (Washington, D.C. Center of Military History United States Army, 1988), pp57

[44] Clarke, Jeffrey J., *Advice and support: The Final Years, The US Army In Vietnam,* (Washington, D.C. Center of Military History United States Army, 1988), pp58

[45] Clarke, Jeffrey J., *Advice and support: The Final Years, The US Army In Vietnam, (*Washington, D.C. Center of Military History United States Army, 1988), pp59

[46] Clarke, Jeffrey J., *Advice and support: The Final Years, The US Army In Vietnam,* (Washington, D.C. Center of Military History United States Army, 1988), pp60

[47] Clarke, Jeffrey J., *Advice and support: The Final Years, The US Army In Vietnam,* (Washington, D.C. Center of Military History United States Army, 1988), pp60

[48] Clarke, Jeffrey J., *Advice and support: The Final Years, The US Army In Vietnam,* (Washington, D.C. Center of Military History United States Army, 1988), pp62

[49] Clarke, Jeffrey J., Advice and support: The Final Years, The US Army In Vietnam, (Washington, D.C. Center of Military History United States Army, 1988), pp64

[50] Clarke, Jeffrey J., Advice and support: The Final Years, The US Army In Vietnam, (Washington, D.C. Center of Military History United States Army, 1988), pp65

[51] Maitland, Terrence, Weiss Stephen, "Raising the Stakes, The Vietnam Experience, (Boston, MA Boston Publishing Company), pp9

[52] Maitland, Terrence, Weiss Stephen, "Raising the Stakes, The Vietnam Experience, (Boston, MA Boston Publishing Company), pp10-11

[53] Maitland, Terrence, Weiss Stephen, "Raising the Stakes, The Vietnam Experience, (Boston, MA Boston Publishing Company), pp11

[54] Maitland, Terrence, Weiss Stephen, "Raising the Stakes, The Vietnam Experience, (Boston, MA Boston Publishing Company), pp11

[55] Clarke, Jeffrey J., Advice and support: The Final Years, The US Army In Vietnam, (Washington, D.C. Center of Military History United States Army, 1988), pp63

[56] Clarke, Jeffrey J., Advice and support: The Final Years, The US Army In Vietnam, (Washington, D.C. Center of Military History United States Army, 1988), pp63-64

[57] Stanton, Shelby, US Army Uniforms of the Vietnam War,(Harrisburg, PA, Stackpole Books,1988), pp 76

[58] Stanton, Shelby, US Army Uniforms of the Cold War1948-1973, (Mechanicsburg, PA, Stackpole Books, 1994), pp 130

[59] Stanton, Shelby; US Army Uniforms of the Vietnam War; (Harrisburg, PA, Stackpole Books, 1988), pp 225

[60] Stanton, Shelby, US Army Uniforms of the Cold War1948-1973, (Mechanicsburg, PA, Stackpole Books, 1994), pp 117

[61] MACV OBSERVER, VOLUME IV, No. 47; MARCH 26, 1966.

[62] Stanton, Shelby, US Army Uniforms of the Vietnam War, (Harrisburg, PA, Stackpole Books, 1988), pp 62, 65

[63] Stanton, Shelby U.S. Army Uniforms of the Vietnam War, (Harrisburg, PA, Stackpole Books, 1988), pp 65

[64] Clarke, Jeffrey J., Advice and support: The Final Years, The US Army In Vietnam, (Washington, D.C. Center of Military History United States Army, 1988), pp61

[65] Clarke, Jeffrey J., Advice and support: The Final Years, The US Army In Vietnam, (Washington, D.C. Center of Military History United States Army, 1988),pp62

[66] MACV OBSERVER, 15 JUNE 1968, Courtesy Carl Farren, interview.

[67] Clarke, Jeffrey J., Advice and support: The Final Years, The US Army In Vietnam, (Washington, D.C. Center of Military History United States Army, 1988),pp68

[68] Stanton Shelby; US Army Uniforms of the Vietnam War; (Harrisburg, PA, Stackpole Books, 1988), pp 151

[69] Stanton, Shelby, US Army Uniforms of the Vietnam War, (Harrisburg, PA, Stackpole Books, 1988), pp159

[70] Stanton, Shelby, US Army Uniforms of the Vietnam War, (Harrisburg, PA, Stackpole Books, 1988), pp172

[71] Clarke, Jeffrey J., Advice and support: The Final Years, The US Army In Vietnam, (Washington, D.C. Center of Military History United States Army, 1988),pp215

[72] Clarke, Jeffrey J., Advice and support: The Final Years, The US Army In Vietnam, (Washington, D.C. Center of Military History United States Army, 1988),pp236

[73] Clarke, Jeffrey J., Advice and support: The Final Years, The US Army In Vietnam, (Washington, D.C. Center of Military History United States Army, 1988),pp237

[74] Clarke, Jeffrey J., Advice and support: The Final Years, The US Army In Vietnam, (Washington, D.C. Center of Military History United States Army, 1988),pp238

[75] Clarke, Jeffrey J., Advice and support: The Final Years, The US Army In Vietnam, (Washington, D.C. Center of Military History United States Army, 1988),pp239-240

[76] Lyels, Kevin, Vietnam: US Uniforms in Colour Photographs, (London W1, Windrow & Greene, 1992), pp 53

[77] Lyles, Kevin US Airborne Vietnam, (Hong Kong, Concord Publications Company) pp9

[78] Martin, CSM Mike, Angles in Red Hats; Paratroopers of the Second Indochina War, (Prospect, KY, Harmony House Publishers) pp18

[79] Martin, CSM Mike, Angles in Red Hats; Paratroopers of the Second Indochina War, (Prospect, KY, Harmony House Publishers) pp18

[80] Martin, CSM Mike, Angles in Red Hats; Paratroopers of the Second Indochina War, (Prospect, KY, Harmony House Publishers) pp11

[81] Martin, CSM Mike, Angles in Red Hats; Paratroopers of the Second Indochina War, (Prospect, KY, Harmony House Publishers) pp87

[82] Martin, CSM Mike, Angles in Red Hats; Paratroopers of the Second Indochina War, (Prospect, KY, Harmony House Publishers) pp11

[83] Stanton, Shelby, US Army Uniforms of the Cold War1948-1973, (Mechanicsburg, PA, Stackpole Books, 1994), pp 105

[84] Stanton, Shelby, US Army Uniforms of the Cold War1948-1973, (Mechanicsburg, PA, Stackpole Books, 1994), pp 105

[85] Stanton, Shelby, US Army Uniforms of the Cold War1948-1973, (Mechanicsburg, PA, Stackpole Books, 1994), pp 105

[86] Martin, CSM Mike, Angles in Red Hats; Paratroopers of the Second Indochina War, (Prospect, KY, Harmony House Publishers) pp48

[87] Martin, CSM Mike, Angles in Red Hats; Paratroopers of the Second Indochina War, (Prospect, KY, Harmony House Publishers) pp11

[88] Martin, CSM Mike, Angles in Red Hats; Paratroopers of the Second Indochina War, (Prospect, KY, Harmony House Publishers) pp113

[89] Martin, CSM Mike, Angles in Red Hats; Paratroopers of the Second Indochina War, (Prospect, KY, Harmony House Publishers) pp18

[90] Martin, CSM Mike, Angles in Red Hats; Paratroopers of the Second Indochina War, (Prospect, KY, Harmony House Publishers) pp73

[91] Stanton, Shelby, US Army Uniforms of the Cold War1948-1973, (Mechanicsburg, PA, Stackpole Books, 1994), pp 189-191

[92] Stanton, Shelby, US Army Uniforms of the Cold War1948-1973, (Mechanicsburg, PA, Stackpole Books, 1994), pp 189-191

[93] Lyles, Kevin US Airborne Vietnam, (Hong Kong, Concord Publications Company) pp9

[94] Lyles, Kevin US Airborne Vietnam, (Hong Kong, Concord Publications Company) pp11

[95] Martin, CSM Mike, Angles in Red Hats; Paratroopers of the Second Indochina War, (Prospect, KY, Harmony House Publishers) pp140

[96] Martin, CSM Mike, Angles in Red Hats; Paratroopers of the Second Indochina War, (Prospect, KY, Harmony House Publishers) pp18-19

[97] Martin, CSM Mike, Angles in Red Hats; Paratroopers of the Second Indochina War, (Prospect, KY, Harmony House Publishers) pp19

[98] Martin, CSM Mike, LTC Valentine, M. The Black Tigers. Indiana: Walsworth Publishing Company., 1998, pp91-94

[99] Martin, CSM Mike, LTC Valentine, M. The Black Tigers. Indiana: Walsworth Publishing Company., 1998. Pp142-143

[100] Stanton, Shelby, US Army Uniforms of the Vietnam War,(Harrisburg, PA, Stackpole Books,1988), pp 76

[101] Stanton, Shelby, US Army Uniforms of the Cold War1948-1973, (Mechanicsburg, PA, Stackpole Books, 1994), pp 130

[102] Stanton, Shelby; US Army Uniforms of the Vietnam War; (Harrisburg, PA, Stackpole Books, 1988), pp 225

[103] Stanton, Shelby U.S. Army Uniforms of the Vietnam War, (Harrisburg, PA, Stackpole Books, 1988), pp39

[104] Sutherland, Ian D. 1952/1982 Special Forces of the United States Army. San Jose: Bender Publishing., 1982, pp685

[105] Sutherland, Ian D. 1952/1982 Special Forces of the United States Army. San Jose: Bender Publishing., 1982., pp681

[106] Martin, CSM Mike, Angles in Red Hats; Paratroopers of the Second Indochina War, (Prospect, KY, Harmony House Publishers) pp35

[107] Martin, CSM Mike, LTC Valentine, M. The Black Tigers. Indiana: Walsworth Publishing Company., 1998. Pp43

[108] Martin, CSM Mike, LTC Valentine, M. The Black Tigers. Indiana: Walsworth Publishing Company., 1998. Pp43

[109] Stanton, Shelby U.S. Army Uniforms of the Vietnam War, (Harrisburg, PA, Stackpole Books, 1988), pp196

[110] Rottman, Rangers, 1943-1988

[111] Martin, CSM Mike, LTC Valentine, M. The Black Tigers. Indiana: Walsworth Publishing Company., 1998. Pp22

[112] Martin, CSM Mike, LTC Valentine, M. The Black Tigers. Indiana: Walsworth Publishing Company., 1998. Pp22

[113] Stanton, Shelby U.S. Army Uniforms of the Vietnam War, (Harrisburg, PA, Stackpole Books, 1988), pp196

[114] Stanton, Shelby, US Army Uniforms of the Cold War1948-1973, (Mechanicsburg, PA, Stackpole Books, 1994), pp 154

[115] Stanton, Shelby U.S. Army Uniforms of the Vietnam War, (Harrisburg, PA, Stackpole Books, 1988), pp65

[116] Stanton, Shelby, US Army Uniforms of the Cold War1948-1973, (Mechanicsburg, PA, Stackpole Books, 1994), pp 113

[117] Stanton, Shelby, US Army Uniforms of the Cold War1948-1973, (Mechanicsburg, PA, Stackpole Books, 1994), pp 113-114

[118] Stanton, Shelby, US Army Uniforms of the Cold War1948-1973, (Mechanicsburg, PA, Stackpole Books, 1994), pp 122

[119] Sutherland, Ian D. 1952/1982 Special Forces of the United States Army. (San Jose: Bender Publishing.),1982., pp668

[120] Stanton, Shelby U.S. Army Uniforms of the Vietnam War, (Harrisburg, PA, Stackpole Books, 1988), pp117

[121] Stanton, Shelby U.S. Army Uniforms of the Vietnam War, (Harrisburg, PA, Stackpole Books, 1988), pp65

[122] Stanton, Shelby U.S. Army Uniforms of the Vietnam War, (Harrisburg, PA, Stackpole Books, 1988), pp9-10

[123] Stanton, Shelby; Rangers at War. New York: Ivy Books, 1992. Pp259-260

[124] Lyles, Kevin US Airborne Vietnam, (Hong Kong, Concord Publications Company) pp9

[125] Cutler, Thomas J. *Brown Water, Black Beret.* Annapolis, Maryland: Naval Institute Press, 1988.pp16

[126] Cutler, Thomas J. *Brown Water, Black Beret.* Annapolis, Maryland: Naval Institute Press, 1988.pp20

[127] Cutler, Thomas J. *Brown Water, Black Beret.* Annapolis, Maryland: Naval Institute Press, 1988.pp20

[128] Cutler, Thomas J. *Brown Water, Black Beret.* Annapolis, Maryland: Naval Institute Press, 1988.pp20

[129] Cutler, Thomas J. *Brown Water, Black Beret.* Annapolis, Maryland: Naval Institute Press, 1988.pp21

[130] Cutler, Thomas J. *Brown Water, Black Beret.* Annapolis, Maryland: Naval Institute Press, 1988.pp21

[131] Cutler, Thomas J. *Brown Water, Black Beret.* Annapolis, Maryland: Naval Institute Press, 1988.pp21

[132] Cutler, Thomas J. *Brown Water, Black Beret.* Annapolis, Maryland: Naval Institute Press, 1988.pp22
[133] Cutler, Thomas J. *Brown Water, Black Beret.* Annapolis, Maryland: Naval Institute Press, 1988.pp24
[134] Cutler, Thomas J. *Brown Water, Black Beret.* Annapolis, Maryland: Naval Institute Press, 1988.pp25
[135] Cutler, Thomas J. *Brown Water, Black Beret.* Annapolis, Maryland: Naval Institute Press, 1988.pp28
[136] United States Navy and Vietnam Conflict; Washington Naval History Division, Dept. of the Navy; 1976-1977. pp305
[137] United States Navy and Vietnam Conflict, Washington Naval History Division, Dept. of the Navy; 1976-1977.pp306
[138] United States Navy and Vietnam Conflict; Washington Naval History Division, Dept. of the Navy; 1976-1977., pp308
[139] United States Navy and Vietnam Conflict; Washington Naval History Division, Dept. of the Navy; 1976-1977., pp308
[140] National Geographic, pp275
[141] National Geographic, pp275
[142] National Geographic, pp277
[143] National Geographic, pp278
[144] United States Navy and Vietnam Conflict; Washington Naval History Division, Dept. of the Navy; 1976-1977., pp309
[145] National Geographic, pp287
[146] United States Navy and Vietnam Conflict; Washington Naval History Division, Dept. of the Navy; 1976-1977., pp309
[147] United States Navy and Vietnam Conflict; Washington Naval History Division, Dept. of the Navy; 1976-1977., pp313
[148] United States Navy and Vietnam Conflict, Washington Naval History Division, Dept. of the Navy; 1976-1977.pp314
[149] Stanton, pp156
[150] Cutler, Thomas J. *Brown Water, Black Beret.* Annapolis, Maryland: Naval Institute Press, 1988.pp43
[151] United States Navy and Vietnam Conflict, Washington Naval History Division, Dept. of the Navy; 1976-1977. pp 216
[152] Stanton, Shelby, US Army Uniforms of the Cold War1948-1973, (Mechanicsburg, PA, Stackpole Books, 1994), pp138
[153] Cutler, Thomas J. *Brown Water, Black Beret.* Annapolis, Maryland: Naval Institute Press, 1988.pp35
[154] Cutler, Thomas J. *Brown Water, Black Beret.* Annapolis, Maryland: Naval Institute Press, 1988.pp34
[155] Cutler, Thomas J. *Brown Water, Black Beret.* Annapolis, Maryland: Naval Institute Press, 1988.pp34
[156] United States Navy and Vietnam Conflict; Washington Naval History Division, Dept. of the Navy; 1976-1977. pp336
[157] Lyles, pp15 Airborne
[158] Lyels, Kevin, Vietnam: US Uniforms in Colour Photographs, (London W1, Windrow & Greene, 1992),pp15
[159] Stars and Stripes, May,1969
[160] Pfiester, Sam lee, Perfect War, pp13, 14, and 20.
[161] Pfiester, Sam lee, Perfect War, pp22
[162] Pfiester, Sam lee, Perfect War, pp17, 18, and 19
[163] Shulimson, Blasiol, Smith, and Dawson US Marines In Vietnam The Defining Year; 1968, (Washington, DC., History and Museum Division HQ, US Marine Corps, 1997), pp23
[164] Shulimson, Blasiol, Smith, and Dawson US Marines In Vietnam The Defining Year; 1968, (Washington, DC., History and Museum Division HQ, US Marine Corps, 1997), pp64
[165] Shulimson, Blasiol, Smith, and Dawson US Marines In Vietnam The Defining Year; 1968, (Washington, DC., History and Museum Division HQ, US Marine Corps, 1997), pp645
USMC Command History, pp23
[166] Shulimson, Blasiol, Smith, and Dawson US Marines In Vietnam The Defining Year; 1968, (Washington, DC., History and Museum Division HQ, US Marine Corps, 1997), pp23
[167] Stanton, Shelby U.S. Army Uniforms of the Vietnam War, (Harrisburg, PA, Stackpole Books, 1988), pp8
[168] Shulimson, Blasiol, Smith, and Dawson US Marines In Vietnam The Defining Year; 1968, (Washington, DC., History and Museum Division HQ, US Marine Corps, 1997), pp645
[169] Shulimson, Blasiol, Smith, and Dawson US Marines In Vietnam The Defining Year; 1968, (Washington, DC., History and Museum Division HQ, US Marine Corps, 1997), pp645
[170] Shulimson, Blasiol, Smith, and Dawson US Marines In Vietnam The Defining Year; 1968, (Washington, DC., History and Museum Division HQ, US Marine Corps, 1997), pp645
[171] Shulimson, Blasiol, Smith, and Dawson US Marines In Vietnam The Defining Year; 1968, (Washington, DC., History and Museum Division HQ, US Marine Corps, 1997), pp645
[172] Shulimson, Blasiol, Smith, and Dawson US Marines In Vietnam The Defining Year; 1968, (Washington, DC., History and Museum Division HQ, US Marine Corps, 1997), pp645
[173] Shulimson, Blasiol, Smith, and Dawson US Marines In Vietnam The Defining Year; 1968, (Washington, DC., History and Museum Division HQ, US Marine Corps, 1997), pp367
[174] Leatherneck Magazine, June 1967
[175] Lyels, Kevin, Vietnam: US Uniforms in Colour Photographs, (London W1, Windrow & Greene, 1992), pp53
[176] Shulimson, Blasiol, Smith, and Dawson US Marines In Vietnam The Defining Year; 1968, (Washington, DC., History and Museum Division HQ, US Marine Corps, 1997), pp649
[177] Shulimson, Blasiol, Smith, and Dawson US Marines In Vietnam The Defining Year; 1968, (Washington, DC., History and Museum Division HQ, US Marine Corps, 1997), pp649
USMC Command History, pp19
[178] Shulimson, Blasiol, Smith, and Dawson US Marines In Vietnam The Defining Year; 1968, (Washington, DC., History and Museum Division HQ, US Marine Corps, 1997), pp649
[179] Shulimson, Blasiol, Smith, and Dawson US Marines In Vietnam The Defining Year; 1968, (Washington, DC., History and Museum Division HQ, US Marine Corps, 1997), pp19
[180] Shulimson, Blasiol, Smith, and Dawson US Marines In Vietnam The Defining Year; 1968, (Washington, DC., History and Museum Division HQ, US Marine Corps, 1997), pp649